ISBN 978-0-282-75293-4
PIBN 10863303

1 MONTH OF
FREE
READING

at
www.ForgottenBooks.com

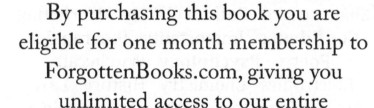

By purchasing this book you are eligible for one month membership to ForgottenBooks.com, giving you unlimited access to our entire collection of over 1,000,000 titles via our web site and mobile apps.

To claim your free month visit:
www.forgottenbooks.com/free863303

WILD FLOWERS.

6058
13

BY ANNE PRATT,

AUTHOR OF

COMMON THINGS OF THE SEA-SIDE, OUR NATIVE SONGSTERS,

ETC. ETC.

" They will bear them a thought of the sunny hours,
And the dreams of their childhood—Bring flowers, Wild Flowers ! "

PUBLISHED UNDER THE DIRECTION OF
THE COMMITTEE OF GENERAL LITERATURE AND EDUCATION,
APPOINTED BY THE SOCIETY FOR PROMOTING
CHRISTIAN KNOWLEDGE.

VOL. II.

LONDON:

PRINTED FOR THE

SOCIETY FOR PROMOTING CHRISTIAN KNOWLEDGE;

SOLD AT THE DEPOSITORY,
GREAT QUEEN STREET, LINCOLN'S INN FIELDS;
4, ROYAL EXCHANGE; 16, HANOVER STREET, HANOVER SQUARE;
AND BY ALL BOOKSELLERS.
1853.

LONDON :
R. CLAY, PRINTER, BREAD STREET HILL.

PREFACE.

THE reception which the first volume of this little book has met with from the public, induces the Author to hope that a further volume may be welcome also. Never was there a time when the direction of our Saviour

to " Consider the lilies" was more willingly
followed than now; and knowing well that
the love of Nature is a great means alike of
mental improvement and of happiness, the
Author rejoices in contributing in any way to
its diffusion. In endeavouring to illustrate,
both by pen and pencil, some of our commonest
Wild Flowers, she has had much to render the
occupation agreeable. It is in itself a pleasant
toil; and while she has been cheered, on the
one hand, by the approval of the highest Lady
of the land—our beloved and revered Queen—
it has been no small gratification to know, that
some in lowliest life have read these simple
details with profit. More than one case has

PREFACE.

been made known to her, in which some have gone forth into the fields at the close of the day of toil, with this book in hand, and taught to their wives and little ones the names and uses of many of those flowers which hitherto they had, perhaps, classed under the general name of weeds. If they have gained one new idea of grace and beauty, or have learned one lesson of the great truth, that "God is love," this little book will not have been written in vain.

Dovor,
   March, 1853.

CONTENTS.

DANDELION.

Wild Flowers.

DANDELION.—*Leontodon Taraxacum.*
Class Syngenesia. *Order* Æqualis. *Nat. Ord.* Compositæ.
Compound Flowers.

Thousands of these "Sunflowers of the spring" glitter in April on grassy plains and slopes, nor will even the chilling frosts of Christmas strip the meads of every bright Dandelion flower.

The leaves of this plant are, on the Continent, much eaten as salad, and are sold in many markets. On one occasion, when a swarm of locusts had destroyed the harvest in Minorca, the inhabitants of that island were compelled to subsist for a time almost wholly on such food as they could gather in the fields, and these plants were then their chief support. The roots are ground, and mixed instead of chicory with coffee; but they are chiefly valuable as a remedy in long-standing liver complaints. The root is for this purpose sliced and boiled, and the decoction thus made will, if persevered with as a medicine, prove an excellent tonic, and will clear the complexion far more surely than the most renowned cosmetic can do.

The leaves of the Dandelion all grow direct
from the root. They are deeply cut, and the
lobes turn backward so as to have suggested
both the scientific and familiar names of the
plant. The former is from the Greek *leon*, a lion,
and *odous, odontos*, tooth; and the English name
is but a corruption of the French *Dent de Lion*.
The flower-stalk is tubular. The ball of down
which succeeds the flower, "the schoolboy's
clock in every town," is too remarkable to have
escaped the notice of any, and most children
would understand the allusion of the poet,—

"Then did we question of the down-balls, blowing
  To know if some slight wish would come to pass;
If storms we fear'd, we sought where there was blowing
  Some meadow flower, which was our weather-glass."

Coles, in his Introduction to the Knowledge
of Plants, says, " If the down flyeth off Colts-
foot, Dandelion, and Thistle, when there is no
winde, it is a signe of rain." The abundance
of this seed renders the Dandelion very general,
and it is the more difficult of extirpation, as
every inch of root will form buds and fibres, and
thus produce a new plant. Our domestic animals
all leave it untouched on the pasture land.

EARLY PURPLE ORCHIS.—*Orchis mascula.*

Class GYNANDRIA. *Order* MONANDRIA. *Nat. Ord.* ORCHIDACEÆ. ORCHIDEOUS TRIBE.

BY the end of May several of our beautiful native Orchis plants are in flower, and we may find this by the middle of the month. It is, too, a very frequent plant, and, like its companion the Blue-bell, is to be found in almost every wood and on many a hedgebank. It has a very succulent stem about a foot high, around which some leaves clasp, and the remainder grow at the root, and are very conspicuous by their large dark purple spots. The flower is strongly scented, and the odour, at no time pleasant, becomes during the evening so powerfully disagreeable, that few can bear the plant in a room. The blossoms are of different shades of purple, spotted with a darker tint. The root, which is composed of two tubers, contains a nutritious flour-like substance, from which the salep formerly so much in use was made. Salep is little used now in our country, but the warm bason of salep, and the old " saloop house," were once as well-known as the cup of Mocha, and the modern coffee-house.

We have ten species of the genus Orchis, besides many orchideous plants which are separated into other genera. The largest of our Orchises is one commonly called the Lady Orchis, having about as much resemblance to a lady as the little figures cut out of paper by children. It is the Great Brown-winged Orchis (*Orchis fusca*), a very beautiful flower of our chalky pastures and woods. The author once gathered a specimen from a wood in Kent, which was two feet and some inches in height, with a close spike of flowers as large as a bunch of grapes. The smallest species of the genus is the pretty Dwarf Dark-winged Orchis (*Orchis ustulata*), which is not above four or five inches in height, and its petals are of so dark a brownish-purple, that the flower looks as if it had been scorched by a flame. It is not un-common in English meadows where the soil is of chalk, and it blossoms in June.

BLACKTHORN.

BLACKTHORN, or SLOE.—*Prunus spinosa.*

Class Icosandria. *Order* Monogynia. *Nat. Ord.* Rosaceæ.
Rose Tribe.

EVERY country child knows the bush so frequent in the hedges, and so prized in autumn for its small harsh fruits, which have much the flavour of an unripe damson. Bloom-field describes the Farmer's Boy, in expectation of some young companion who has promised to come and relieve his solitude, as placing the bough with its sloes over the fire, that the roasted fruits may be his reward. Assuredly, the Sloe would need some such improvement, were it to be eaten by older persons, yet we have not despised it, when in early days we

"Rambled in the field
To gather austere berries from the bush,
Or search the coppice for the clustering nuts."

The flavour of the Sloe is much improved by being placed in bottles underground till winter, when it forms a very pleasant preserve, though, perhaps, not a very wholesome one.

The old name of Blackthorn is appropriate, for the tree is thorny, and the dark colour of its bark is made the more remarkable by the snowy whiteness of the flowers. Country people will tell you, too, that its name alludes to the fact, that this blossom appears during the " black winds of March," or " Blackthorn winter." They are in full perfection before a single leaf is on the spray, though in the month of April, both leaves and flowers may be seen on the bough.

> " All other trees are wont to wear
>   First leaves, then flowers, and last
> Their burthen of rich fruit to bear,
>   When summer's pride is past ;
> But thou, so prompt thy flowers to show,
>   Bear'st but the harsh unwelcome sloe."

This circumstance it is, which chiefly distin-guishes the Sloe from the Wild Bullace-tree (*Prunus insititia*), for the latter plant, though much resembling it, is in full leaf before its flowers appear, and does not blossom until May.

The thorns of the Sloe-bush are very strong and sharp. The fruits are used to adulterate port wine, and the leaves are said to be mixed with tea.

Just at the season when the white blossoms are gleaming on our every hedge-bank, the hills and vales of Palestine are made gay with the beau-tiful flowers of the Sloe and the Almond-tree.

LESSER PERIWINKLE.—*Vinca minor.*

Class PENTANDRIA. *Order* MONOGYNIA. *Nat. Ord.* APOCYNEÆ. PERIWINKLE TRIBE.

THE rich blue flower and glossy green leaves of the Periwinkle are very beautiful, and are seen to advantage among the delicate Anemones and wan Primroses near which it grows. The plant is found more frequently in woods rendered moist by streams than elsewhere, but we can hardly call it a common flower, though in many woodland spots it is abundant. In Kent it is not uncommon, and in some of our western counties it is plentiful. In Devonshire it often covers large spaces of ground with its leaves and blossoms. The stem is trailing, and the wood of the shoots is remarkably tough, and so cord-like, that from the stem the genus received its name, from *vincio,* to bind. The Dutch call it *Sinn-green,* or evergreen, as its leaves are to be seen throughout the year. The Italians bind down the grassy sod with its shoots, and from this association with the tomb, they term it the Flower of Death. The pistil of the flower is clothed with minute hairs, and is a very beautiful object.

Hurdis describes this graceful plant when growing in the rural garden,—

"See where the sky-blue Periwinkle climbs
  E'en to the cottage eaves, and hides the wall,
And dairy lattice, with a thousand eyes,
  Pentagonally form'd, to mock the skill
Of proud geometers."

The Larger Periwinkle (*Vinca major*), which is so common in gardens and shrubberies, is sometimes found in the hedges, but is a doubtful native, being seen mostly near villages, where it may be the outcast of the garden. The stem of this species is nearly erect, and both flowers and leaves are larger than those of the kind represented in the engraving. A white variety of this small species is sometimes found, especially in Devonshire, and the author has found it about Boughton, near Canterbury. A variety with double flowers and leaves variegated with straw-colour, is also frequent in gardens. Both species are in blossom during May and June, and are perennial plants. Their properties are acid and astringent.

GREATER STITCHWORT.—*Stellaria Holostea.*

Class DECANDRIA. *Order* TRIGYNIA. *Nat. Ord.* CARYOPHYLLEÆ. CHICKWEED TRIBE.

FEW flowers delight us more by their beauty than this pearly Stitchwort, which blossoms by our waysides and among the primroses of the wood, in the month of May. It was called in early times White-flowered grass, and is now known in country places as the Satin flower, and Adder's meat. The delicate green leaves are almost as beautiful as the blossoms, and tufts of these may be seen even before the spring has arrived. The whole plant is very brittle. Gerarde says, " It is called in Latin *Iota ossa*, in English, All-bones; whereof I see no reason except it be by the figure *Autonomia*, as when we say in English, He is an honest man, our meaning is that he is a knave; for this is a tender hearbe, having no such bony substance."

This genus is named from *Stella*, a star, all the eight species having white flowers of starry form. The Chickweed (*Stellaria media*), which is given to caged birds, and the young buds and seeds of which form so valuable a supply of food to our wild songsters, is the most frequent kind, flowering and ripening its seeds throughout the greater part of the year. Another common species is the Lesser Stitchwort (*Stellaria graminea*), which has much smaller flowers than those represented on our pages, and is altogether a more slender plant. It grows on dry heaths and pastures, and may readily be known by the more deeply cloven petals. The glaucous Marsh Stitchwort (*Stellaria glauca*) has blossoms nearly as large as those of the Satin flower, but they grow singly, instead of being in clusters: and the Bog Stitchwort (*Stellaria uliginosa*), a plant of our bogs and ditches, has very minute white blossoms in loose clusters, with broader leaves.

There are besides, the Wood Stitchwort (*Stellaria nemorum*), common among trees and bushes, with heart-shaped leaves; the lowly Alpine Stitchwort (*Stellaria cerastoides*), a flower of the Scottish mountains; and the rare Many-stalked Stitchwort (*Stellaria scapigera*), which is found only on the borders of Loch Nevis, and on the hills north of Dunkeld.

WATER VIOLET.

WATER VIOLET.—*Hottonia palustris.*

Class PENTANDRIA.   *Order* MONOGYNIA.   *Nat. Ord.* PRIMULACEÆ.
PRIMROSE TRIBE.

SOME of our still waters, especially those having a gravelly soil, are made very beautiful in June by the flower of the Water Violet. It is always pleasant, when rambling in the country, to find the waters lying amid grass and flowers :

"Green tufted islands casting their soft shades
　Across the lakes; sequester'd leafy glades,
　That through the dimness of their twilight, show
　Large Dock-leaves, spiral Foxgloves, or the glow
　Of the wild Cat's Eyes, or the silvery stems
　Of delicate Birch-trees, or long grass which hems
　A little brook."

Keats probably meant the Herb Robert by the name "Cat's Eye," as that is one of its names in country places.

Poets have almost left our Water Violet unnoticed, for though frequent in some districts of England, and highly ornamental to ponds and ditches, yet it can hardly be called a common flower. It is in Ireland a very rare plant, and quite unknown in Scotland. As our engraving will show, it in no way resembles the violet, nor has it the odour of that sweet flower, the blossoms growing in whorls around the stem from six to ten in number, and the colour being more like that of the Cuckoo

flower, which in early spring is found in moist
meadows and woods, and which is of a pale
lilac. The leaves grow around the root, and are
so plume-like as to render another name of the
plant more significant. Thus it is called Fea-
ther Foil, and is also known in various places
a Water Milfoil, and Water Gilly-flower. Its
name of Hottonia was given by Boerhave, in
honour of Dr. Peter Hotton, curator of the
Leyden Botanic Garden, who died in 1709.

The leaves of this plant are all submersed,
affording shelter to many water insects and
shell-fish; and the root fibres run far into the
soil, while the blossom rises high above the
surface of the pond. The root is perennial,
and from its crown spring several leafy run-
ners, which taking root at their extremities,
produce flowers in the following summer.

We have but one native species, though a
variety has been found in Northamptonshire,
bearing a red flower. The Water Violet is
not difficult of culture, and as it is so ornamental to the waters, it is to be regretted that
it is not oftener planted. The ripe seeds
thrown into the ponds in one summer, will
produce flowers in the following season.

COMMON KIDNEY VETCH.—*Anthyllis vulneraria.*

Class DIADELPHIA.   *Order* DECANDRIA.   *Nat. Ord.* LEGUMINOSÆ.
PEA AND BEAN TRIBE.

THIS handsome flower is common on dry pastures and hills, and very general on seaside cliffs.   On the chalky heights of Dovor it is most abundant, covering them as early as May with its yellow blossoms, and blooming on till August or September.   On these cliffs it is very luxuriant, and always yellow; but the Rev. C. A. Johns finds that it has, on the Cornish coast, a stunted habit of growth, and it bears there, as well as on some other places, flowers of a crimson, purple, cream-coloured, or white hue.   Linnæus remarked that in Œland, where the soil is a red calcareous clay, the flowers are red; but in Gothland, where the soil is white, the flowers are white.   It is a common flower on almost all parts of the continent, and in Portugal is usually of a rich red hue.

The blossoms of the Kidney Vetch are crowded into heads, which grow two together, at the end of each stalk, and the plant is rendered very peculiar by its clear white swollen flower-cups, which are thickly covered with a short soft wool.   It is in country places called Lamb's Foot, Ladies' Fingers, and Woundwort.

" Along the expanse of lengthening meads were flung,
    Mingled with Lady Smocks and Daisies white,
Lamb's Foot and Speedwell, and the lovely sight
Of Hawthorn blossom, fragrant on the gale."

Gesner first ascribed to this plant its vulnerary properties, but saving that it is downy and soft as lint, these are not very apparent. Baxter quotes Threlkeld, who says that it was regularly sold in his time, 1726, in the markets of Ireland, by the name of staunch, because of its astringent properties.

The plant has been recommended on high authority as affording good pasturage for sheep, and a good yellow dye may be obtained from its petals.   It is the only British species; but some garden kinds obtained from the south of Europe have a much greater quantity of down on the flower-cups.

RED DEAD NETTLE.

RED DEAD NETTLE.—*Lamium purpureum.*

Class DIDYNAMIA. *Order* GYMNOSPERMIA. *Nat. Ord.* LABIATÆ. LABIATE TRIBE.

EVEN while winter is still with us, and before we have begun to look for the wild flowers, the little blossom of this Dead Nettle gladdens us on the hedge-bank. Perhaps, some sunny morning of February tempts us to a walk in the rural lane, and we are there reminded of Clare's description of the joy with which children hail the unexpected gleam.

> " And oft in pleasure's dreams they hie
>   Round homestead by the village side,
> Scratching the hedge-row mosses by,
>   Where painted pooty shells abide:
> Mistaking oft the ivy spray
>   For leaves that come with budding spring,
> And wondering in their search for play,
>   Why birds delay to build and sing."

Even then, however, the children may gather a small nosegay, for winter has a few wild flowers, and our Dead Nettle, with its dull purplish-red blooms, and the Daisy and Dande-lion, and the Blue Ivy-leaved Speedwell, and the Grey Procumbent Speedwell, or Winter

Weed, are the forerunners of the multitudes of spring.

This Dead Nettle is to be found at all parts of Britain, and is in bloom until September. The upper leaves have usually a purplish tinge, and are covered with silky hairs. Linnæus tells us that this plant is commonly boiled in Upland for greens, and old herbalists record a similar use of the Dead Nettle in our own land. It is also in great esteem among country people, as a healing application to wounds.

The genus Lamium received its name from the Greek, Laimos, a throat, from the shape of the flowers. The Labiate or lipped tribe of plants in which it is included, are all wholesome, and most of them are fragrant and aromatic, though the odour of the Dead Nettles is not agreeable. Many very useful plants, as the sage and different kinds of mint, belong to it. They have all square stems, opposite leaves and two-lipped blossoms, and red, purple, and lilac, are their most common colours.

COWSLIP.—*Primula veris.*

Class PENTANDRIA. *Order* MONOGYNIA. *Nat. Ord.* PRIMULACEÆ.
PRIMROSE TRIBE.

THIS flower is in many counties called Paigle. Several rustic practices are connected with it. Some of our old herbalists, who in their zeal for the simples and herb-drink which they recommended, lost no opportunity of declaiming against the "drugs" of the physician, praise very highly the ointment made of Cowslip petals, and one of them says, "Our city dames know well enough that the distilled water of the cowslip adds beauty, or at least restores it when lost." The ointment is still used in villages to remove freckles and sunburn. Then the Cowslip is a source of continual delight to the children, who ingeniously make the flowers up in balls. This is done by picking off the clusters from the tops of the stems, and hanging a number of them across a string stretched between two chairs. Great pains are taken to press the flowers carefully together, and to draw up the string so as to bring them into a ball. The Cowslip gatherer, who has made many of these balls, knows well how important it is to their roundness, that all the flowers should be fully blown, and will search far and wide over the meadows to find the blossoms in perfection. A pleasant and even medicinal wine is made of the Cowslip bell, and the leaves are often boiled for the table; but our beautiful plant is disliked by the agriculturist, for it is not eaten by cattle, and the large leaves occupy much room on the field.

The Cowslip is in blossom during April and May. It is often introduced into gardens, but cultivation soon changes its colour, and it becomes first an orange brown, and afterwards a deep red. Our old poets, as Shakspeare, Milton, and Ben Jonson, have noticed this flower, nor have modern poets overlooked it. The " Freckled Cowslip—"

> " Cinque spotted like the crimson drops
>   I' the bottom of a Cowslip ;"

and

> " Cowslips wan that hang the pensive head ;"

are familiar descriptions. Ben Jonson makes one in the Shepherd's Holiday to exclaim :—

> " Strew, strew the smiling ground
>   With every flower, yet not confound
>   The Primrose drop, the Spring's own spouse,
>   Bright day's eyes and the lips of cows,
>   The garden star, the queen of May,
>   The Rose to crown the holiday."

SEA CABBAGE.—*Brassica oleracea.*

Class TETRADYNAMIA. *Order* SILIQUOSA. *Nat. Ord.* CRUCIFERÆ. CRUCIFEROUS TRIBE.

THIS is a plant peculiar to the sea-shores, and is very abundant on many cliffs there. At Dovor no flower, save the bright blue Viper's Bugloss of summer, is so ornamental to the chalky heights; and from May to August its handsome pale yellow clusters may be seen from the sea-beach far away. Even when the autumnal winds sweep in with wild music among the hills, shaking the long woody stems of the Cabbage plant as they hang out in tufts from the crevices, even then we may find a few blossoms sheltered by some prominence, and lingering in delicate beauty till winter. Nor when all the flowers are gone does the plant cease to lend a grace to the spot. The leaves are richly tinted with dark green, or pale yellow, or with delicate lilac, or rich deep purple, the surface being well covered with that greyish-white powder so common on the leaves of sea-side plants. The leaves of the stem are oblong, but those around the root are waved and fleshy. Though very bitter in their uncooked state, they may, by repeated washings, be rendered fit for food, and they are often boiled and eaten at sea-coast towns. At Dovor they are gathered by boys from the cliffs, and carried about for sale. This plant is the origin of all the several kinds of garden Cabbage. The name of the genus is from the Celtic *Bresic*, Cabbage.

The Isle of Man Cabbage (*Brassica Monensis*) is another sea-side species. It grows on sandy sea-shores on the north-western coasts of Britain, and its bright lemon-coloured blossoms, veined with purple, appear in July. The genus Brassica contains, besides, the Common Wild Navew (*Brassica campestris*), frequent on field borders, and very similar in appearance to the Charlock. The Rape or Coleseed (*Brassica Napus*), which is cultivated for the oil of its seeds, and the Common Turnip (*Brassica Rapa*), are also enumerated among British plants, both being often found on waste places, but they are not truly wild.

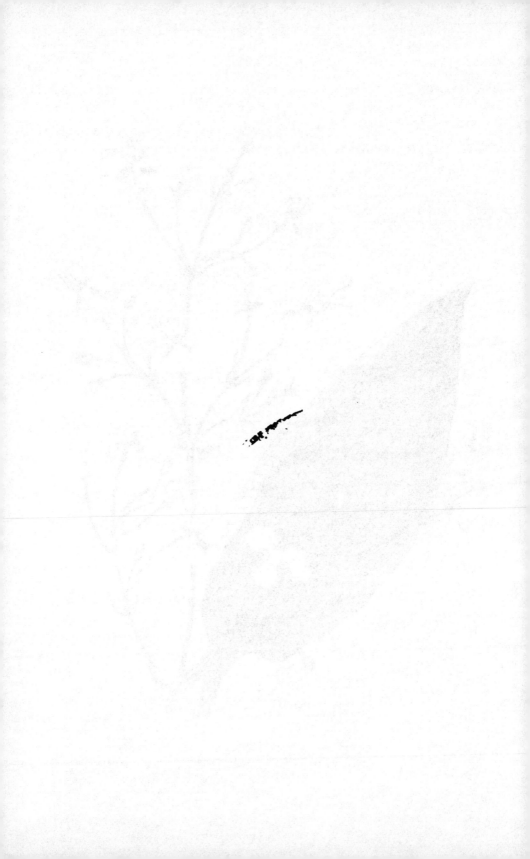

GREAT WATER PLANTAIN.—*Alisma Plantago.*

Class HEXANDRIA.   *Order* POLYGYNIA.   *Nat. Ord.* ALISMACEÆ. WATER-PLANTAIN TRIBE.

THIS plant is made more conspicuous by its large leaves and tall stems, than by its numerous but small blossoms of a delicate lilac hue. It often attains the height of three or four feet, the leaves being all on long stalks, and arising from the root. It is to these thickly-ribbed leaves, so like those of the plantain, that the plant owes its name. It is also known in villages as the Great Thrumwort. The genus is supposed to be so called from the Celtic *alis*, water. It is in blossom from June to August.

The Water Plantain has long been held in repute as a cure for canine madness, and is said still to be regarded as efficacious for this malady over the greater part of the Russian Empire. "We are told," says Dr. George Johnston, " that in the government of Isola it has never failed of a cure for the last twenty-five years." Reduced to powder it is spread over bread and butter, and eaten. Two or three doses are said to be sufficient in most cases to remove the illness, and it is said to cure mad dogs themselves, "but," adds our excellent naturalist, "this also is vanity!" In America it is renowned as a remedy against the bite of the rattlesnake.

The tubers of the Water Plantain, like those of the Water Arrow-Head, contain a nutritious substance, and are eaten by the Kalmuck Tartars. Baron Haller remarks, however, of the leaves and stems, that their acrimonious qualities almost equal those of the Crowfoots, and says that the plant has often proved fatal to kine and other animals. If externally applied it will blister the skin.

We have two other British species. The Lesser Water Plantain (*Alisma ranunculoides*) is not uncommon in peaty bogs. Its flowers are larger and paler, but the whole plant is much smaller than the Great Thrumwort. The floating Water Plantain (*Alisma natans*) is found only on lakes among the mountains in Cumberland and North Wales, and very rarely in Scotland. Its flowers grow singly, and its long-stemmed leaves float on the water.

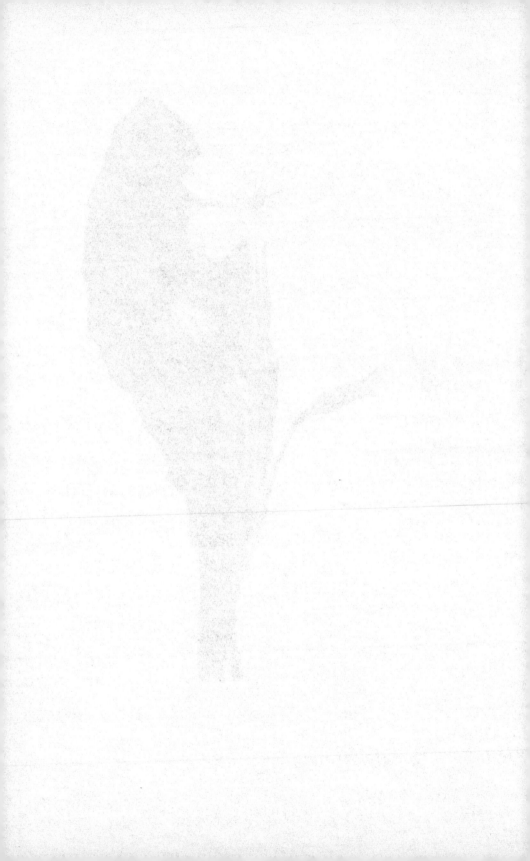

PRIMROSE.—*Primula vulgaris.*

Class PENTANDRIA. *Order* MONOGYNIA. *Nat. Ord.* PRIMULACEÆ. PRIMROSE TRIBE.

OUR primrose well deserves its name (*Prima rosa*, the first rose), for though in the field the daisy may have preceded it, it is the first of the woodland flowers. Even in February,

> " The woodman in his pathway down the wood
> Crushes with hasty feet full many a bud
> Of early primrose ; yet if timely spied,
> Shelter'd some old half-rotten stump beside,
> The sight will cheer his solitary hour,
> And urge his feet, to strive and save the flower."

This blossom is commonly described as sulphur-coloured, but, as the Rev. C. A. Johns has remarked, " the colour of the flower is so peculiar as to have a name of its own ; artists maintaining that primrose-colour is a delicate green." Our old English poet, Edmund Spenser, thus characterises its tint :—

> " A crimson coronet
> With daffodils and damaske roses set ;
> Bay-leaves betweene,
> And primroses greene
> Embellish the sweet violet."

The flower is in full bloom from March till May, in woods and on banks. It is left untouched by cattle, not one of the primrose tribe being relished by any animals save swine. The petals are gathered by country people and made into an ointment, which, though highly prized in villages, cannot be a remedy of any great power.

We have two other wild primroses, besides the Oxlip and the Cowslip, which are also included in the genus. The Oxlip Primrose (*Primula elatior*) is much like a large cowslip, but it is not a common flower. The Bird's Eye Primrose (*Primula farinosa*) is a very lovely plant, with blossoms of a pale lilac-purple. It is not uncommon on mountainous pastures in the North of England. Its leaves, stems, and flower cups, are sprinkled with a fine white powder, and it has a musky odour. The Scottish Primrose (*Primula Scotica*) is much like it, but is a still more beautiful plant, though smaller and stouter in its habit. The last, however, is a rare species, found only in the Orkneys, and in a few places in the North of Scotland. Neither of these lilac primroses is a spring flower, both appearing in July.

FIELD MADDER.—*Sherardia arvensis.*

Class TETRANDRIA. *Order* MONOGYNIA. *Nat. Ord.* RUBIACEÆ. MADDER TRIBE.

OUR little plant, with its small tufts of lilac blossoms, and its whorled leaves, is of so humble a growth as to escape the notice of many even who love wild flowers. Yet the field which the young blades of corn are colouring with tender verdure, or where the Trefoils are thickly clustering, often hides among its produce this humble weed; and, frequent as it is in all cultivated lands, it is rarely absent from those whose soil is of a gravelly nature. The stems all spread around the roots over the ground. The whorls contain about six leaves, and the little clusters of flowers are sometimes almost hidden among the whorl at the end of the stems. It is in bloom from May to August. The name of Sherardia was bestowed upon it in memory of James Sherard, the eminent English botanist, whose valuable herbarium is still preserved at Oxford; though so small a plant, and one of no particular use to man, might seem but little fitted to commemorate a man whose botanical science was much esteemed.

The name of Madder would suggest that our little plants afforded the dye well known to be yielded by the roots of a plant of that name. But the Dyer's Madder is the *Rubia tinctoria* of the south of Europe, and our plant only resembles it in the mode of growth; in its whorled leaves and its four-cleft blossoms, and not in the nature of its juices. The natural order *Rubiaceæ*, in which our little flower, as well as the true Madder are included, is one of the largest known tribes of plants, containing more than two thousand eight hundred species. The group to which the Field Madder belongs, characterised by its starry growth of leaves, has by some botanists been separated into a distinct order under the name of *Stellatæ*.

The Field Madder is the only British species of Sherardia, but we have, in gardens, a pretty little yellow flower, the Wall Sherardia, which is a native of Italy.

COMMON WALLFLOWER.—*Cheiranthus Cheiri.*

Class TETRADYNAMIA. *Order* SILIQUOSA. *Nat. Ord.* CRUCIFERÆ. CRUCIFEROUS TRIBE.

OUR wild Wallflower gleaming on the old wall or ruined tower, so early in spring, is welcome alike for its beauty and fragrance. It thrives amid March winds and April showers, and may now and then be found in blossom in colder seasons; hence, Gerarde says of it, " The people in Cheshire doe call it Winter Gillyflower." The specimen from which the illustration was made, was gathered from the cliffs of Dovor ; where, as on the walls about that ancient town, the Wallflower is most plentiful. Moir, as well as many other poets, have sung its praises.

> " In the season of the tulip cup,
>   When blossoms clothe the trees,
> 'Tis sweet to throw the lattice up
>   And scent thee on the breeze:
> The butterfly is then abroad,
>   The bee is on the wing,
> And on the hawthorn by the road
>   The linnets sit and sing."

We have but one British species of this plant; and the garden Wallflowers, in all their hues of brown and yellow, and their double petals, are but varieties of this. We never find it in a wild state, however, with the drooping petals which it has in the garden, and it is usually much more shrubby when uncultured. It remains in blossom more or less, through the summer, delighting in the driest soils. It is a favourite flower throughout the East, growing on ruins and walls there, as with us, and cherished, too, as a garden flower. The name of the genus is derived from the Arabic *Kheyry;* but this word is applied by the Arabs to a genus of red flowers, similar to it in little but fragrance. Like all cruciferous plants this is perfectly wholesome, though too pungent to be used as food for man. It is, however, effectual as a remedy against maladies to which sheep are liable, and is sometimes sown with thyme and parsley on pasture lands. Our forefathers called it Yellow Violet, and the Dutch still term it *Violier,* and the Spaniards *Violette Amarella.*

SILVER WEED.—*Potentilla anserina.*

Class ICOSANDRIA.　*Order* POLYGYNIA.　*Nat. Ord.* ROSACEÆ.
ROSE TRIBE.

THIS little plant is known also in country places by the name of Trailing Tansy, from the similarity in the shape of its leaf to that of the fragrant herb. In olden times it was called *Argentina.* The leaves, which are very numerous, are covered, especially on their under surfaces, and in an early stage of growth, with numerous silky hairs, giving them the silvery hue which suggested one of the familiar names of the plant. The flowers are yellow, and soft like velvet.

"And Silver Weed with yellow flowers,
　　Half hidden by the leaf of grey,
　Bloom'd on the bank of that clear brook,
　　Whose music cheer'd my lonely way."

The plant is very common by road-sides and on moist meadows, and the leaves may be seen there during the greater part of the year; the flowers appear in June and July.

The roots of the Silver Weed are eaten either roasted or boiled by the Scottish peasantry, and children in rural districts of England sometimes lay them over a brisk fire,

and then eat them. They are very small, but
to some of us in childhood they seemed quite
as pleasant as the fruit of the chestnut. Ray
says that in his time the boys called them Moors,
and adds, that in the winter season they dug
them up and ate them. He also observes that
he has seen them rooted up and eagerly
devoured by swine. In seasons of scarcity
they have sometimes served the purposes of
bread, and the poor people of the isles of Tiray
and Col have been, on more than one occasion,
almost entirely indebted to these small roots
for their daily food.

A friend, whose early home was a Highland
manse, has described to the writer how eagerly
this plant was gathered in summer time by
the female part of the household, and steeped in
buttermilk to remove the freckles and brown-
ness which the sun had brought to the fair
cheek. It was an old advice of the herbalist
thus to use the plant. It was to be soaked in
butter-milk nine days, and maidens were promised that if they would " wash their faces
therewith, it would make them looke very
faire and beautifull."

COMMON SOW-THISTLE.—*Sonchus oleraceus.*

Class SYNGENESIA. *Order* ÆQUALIS. *Nat. Ord.* COMPOSITÆ. COMPOUND FLOWERS.

THOSE whose early lives were passed in the country, well know this common plant, which is in great request as food for the tame rabbit, and is equally well relished by wild rabbits and hares. It is often called Milk Thistle from its milky juices, and on the same account the French call it *le Laiteron,* though the true Milk Thistle is a prickly plant with a deep purple flower, and has milky white veins in its leaves. The juice of the Sow-Thistle is very similar in its nature to that of the Dandelion and Succory, but it is not used in medicine. The plant, though a favourite food with many animals, especially the sheep, is disliked by horses. The young and tender leaves of the smooth variety are, when cooked, thought superior in flavour to our garden spinach. The yellow radiate flowers may be seen from June till September, and the plant grows commonly in waste places, as well as among the weeds of the garden.

The Corn Sow-Thistle (*Sonchus arvensis*) blooms rather later in the summer, and bears larger yellow blossoms. It is frequent in corn-fields and in waste places. The tall Marsh Sow-Thistle (*Sonchus palustris*), with numerous large yellow flowers, and hairy flower-cups, grows on marshy lands, but is not a common plant. It is found on the Isle of Ely, about Greenwich and Blackwall, at Croydon, and some other places. The Alpine Sow-Thistle (*Sonchus Alpinus*) is rarer still, and very easily known from all the other species by its blue blossoms. This handsome plant occurs on rocky places, by the sides of rivulets in the Clova mountains. It is three or four feet in height.

Several species of Sow-Thistle are, in other lands, eaten as food, and one, common in North America, the Small-flowered Sow-Thistle (*Sonchus Floridanus*), which the American settlers term the Gall of the Earth, is renowned as a cure for the bite of the rattle-snake. The genus is named from the Greek word *soft*, in allusion to the soft and feeble stems of most of the plants which compose it. Some very handsome species are cultivated in gardens; they have all yellow or blue flowers.

RAGGED ROBIN.—*Lychnis Flos-Cuculi.*

Class DECANDRIA. *Order* PENTAGYNIA. *Nat.Ord.* CARYOPHYLLEÆ. PINK TRIBE.

THIS pretty rose-coloured blossom shares with several others of the same season the name of Cuckoo-flower. Our forefathers watched the various connexions between natural objects with more attention than do men of modern times. The fields and lanes were the pages of their almanac. They had their Cuckoo-flowers, and Swallow-worts, and Wake-Robins; and they never forgot that the flowers and the birds came together to bless the green earth, and to fill the air with life and song. They had their old rustic proverbs, which told that the blossom of the Sloe should be the time for the sowing of the barley, and the bursting of the Alder-boughs the season when the eel should stir from its winter quarters and might be taken. The loud note of the cuckoo was associated with the Wood Sorrel, for that was called Cuckoo's-meat; and the Early Orchis, the Arum, the Wood Anemone, and the Lady's Smock, still in country places have the name of Cuckoo's-flower. The

latter plant (*Cardamine pratensis*) is said by
Gerarde to be the true Cuckoo-flower, and
Shakspeare's " cuckoo buds of yellow hue "
are probably the Buttercup, or the equally
early blossoming Marsh Marigold.

Our Ragged Robin grows abundantly in
many moist meadows, and rises above the
grass so as to be very conspicuous, being one
or two feet high.   It blooms by the latter end
of May and during June, and the petals are so
deeply jagged as to have quite a ragged
appearance.   The upper portion of the stem
is clammy, and the lower somewhat hairy.
This plant is also known in country places
as the Bachelor's Button.   The French call
it *Lychnide*.   We have four native species of
the Lychnis genus, all scentless, but having
pretty rose-coloured flowers, in one species
varying to white.

CORN-COCKLE.—*Agrostemma Githago.*

Class DECANDRIA. *Order* PENTAGYNIA. *Nat. Ord.* CARYOPHYLLEÆ.
PINK TRIBE.

As we have a flower which is called Queen of the Meadows, so the corn land has its regal adornment, Agrostemma, signifying the Crown of the Field. Several of our most showy wild flowers grow among the ripening or ripened corn. The Scarlet Poppy, the rich Blue Succory, or deeper-tinted corn Blue-bottle, the Yellow Marigold, and the Lilac Scabious, are all beautiful, but not one is more so than our rich purple Corn-Cockle. None of these flowers is welcome to the cultivator, least of all is the Cockle. " What hurte it doth among corne ! " says Gerarde ; " the spoyle unto bread, as well in colour, taste, and unwholesomeness, is better known than desired." The handsome black glossy seeds are large and heavy ; and the husks, breaking so fine as to elude the miller's care, fill the corn with black specks. There seems no doubt, too, that the flour mixed with that of the wheat, would, if existing in any great quantity, be very pernicious. M. Malapert, in conjunction with

M. Bonnet, proved the seeds to be poisonous, and these botanists ascribe this to the fact of their containing the soapy principle (*saponine*) which occurs in the seeds both in their unripe and matured states, and which also is found in the root of the plant.

As the seeds are numerous, the Corn-Cockle is abundant on ill cultivated soils, the only means of eradicating it being by pulling it up by the hand before it comes into flower.

Some seeds, called Git, or Gith, were employed by the Romans in cookery, and were, in all probability, those of the Fennel-flower, which are not only black in colour, but aromatic in flavour. Those of the Corn-Cockle resemble them in appearance, though not in properties, and hence the specific name of Githago. This is the only British species. It is in blossom in June and July. The stems are downy, and it is most frequent on dry and gravelly soils. The French term it *La nielle*, and the Dutch name it *Koornvlam*. That very common flower of the garden, called the Rose Campion, or Prick-nose, is the Agrostemma Coronaria, and is a native of Italy.

WHITE OX-EYE.

WHITE OX-EYE.—*Chrysanthemum Leucan-themum.*

Class Syngenesia. *Order* Superflua. *Nat. Ord.* Compositæ. Compound Flowers.

THE large white flowers of the Ox-Eye, or Moon Daisy, as it is more frequently called, are among the commonest blossoms of the meadow. They are very beautiful too, standing in large clusters, or bowing among the tall grasses which are just ready for the mower's scythe. The golden centre of the flower probably gained for it its old name of Daisy goldins, and, like the Common Daisy and the Buttercup, it is attractive to children,—

" From the young day when first their infant hands
    Pluck witless the wild flowers."

Joyous little groups of villagers may often be seen sitting at the cottage door, or on the meadow grass, stringing the large white stars on a thread, or piece of wire, and forming no inapt representation of a military plume, while they are well cautioned by their mothers not to suffer the plant to touch the eye, lest its acrid juice cause pain. Ben Jonson, in enumerating some of his favourite flowers, gives us

a goodly list of those which grow either in
field or garden:—

"Bring Corn-flag, Tulip, and Adonis flower,
    Fair Ox-eye, Goldy-locks, and Columbine,
    Pinks, Goulans, King-cups, and sweet Sops-in-wine;
    Blue Harebells, Paigles, Pansies, Calaminth,
    Flower-gentle, and the fair-hair'd Hyacinth;
    Bring rich Carnations, Flour-de-luces, Lilies."

The Ox-Eye blossoms during May and June
and its roots are perennial. The whole plant
is said to be destructive to fleas and other
insects, hence it is sometimes hung up in
country dwellings. It has little odour till it
is bruised, when its scent is perceptible, and
its leaves have a slightly aromatic flavour
when quite fresh. They have been recommended by physicians as an external remedy
for some diseases.

The only other British species of Chrysan-
themum is the beautiful yellow flower, usually
called Wild Marigold, (figured in an earlier
volume,) which in some parts of our country
is very abundant in corn-fields in June and
July, and which in fields ploughed in summer
blossoms again in the autumnal months. This
species is an annual plant.

MEADOW SWEET, OR DROPWORT.
Spiræa Ulmaria.

Class ICOSANDRIA. *Order* MONOGYNIA. *Nat. Ord.* ROSACEÆ.
ROSE TRIBE.

WELL does our beautiful and fragrant flower deserve its names of Meadow Sweet, and Queen of the Meadows, for few are more graceful, and few have a more powerful odour. Calder Campbell, in his description of the flowers of the stream-side, alludes to the injurious effect of this scent, which, however, is not hurtful when borne on the meadow breeze, though very deleterious in a close room :—

" Bright as the birds of Indian bowers,
　　Whose crimson plumage blent with green
Their emerald leaves and vermeil flowers
　　Resemble, Willow herbs are seen
To nod from banks, from whence depend
　· Rich cymes of fragrant Meadow Sweet;
Alas ! those creamy clusters lend
　　A charm, where death and odour meet."

The Meadow Sweet blossoms in June and July, and its stem is from two to four feet in height. When growing in abundance, as it often does, by the side of the stream, it renders the margins as white as if snow had fallen there. The whole plant is very astringent, but though formerly used medicinally, is not now valued. It is eaten by sheep and swine, but is disliked by most animals. The flowers infused in boiling water impart a fine flavour to it, which is increased by distillation.

The Common Dropwort (*Spiræa Filipendula*) is not so frequent a flower as the Meadow Sweet, though not uncommon on dry chalky or gravelly pastures. It blossoms in July, and has much resemblance to the plant figured on our pages, but the flowers are tipped with rose colour. The root of this plant consists of rather long tubers, which, when dried and reduced to powder, are said to be an excellent substitute for bread flour, and which in times of scarcity would be of no small value.

The only other native species, the Willow-leaved Dropwort (*Spiræa salicifolia*), is a small shrub. It grows in moist woods in several parts of the North of England, in Wales and Scotland, bearing thick close clusters of rose-coloured flowers in July. The name of this genus is of Greek origin. Several very ornamental shrubs belonging to it are cultivated in gardens.

HERB PARIS.—*Paris quadrifolia.*

Class OCTANDRIA. *Order* TETRAGYNIA. *Nat. Ord.* TRILLIACEÆ. HERB PARIS TRIBE.

WE have very few native plants with green blossoms. The Dog's Mercury, it is true, grows on every way-side, and the Black Bryony with its long clusters of small green flowers, hangs about the trees. The Adoxa, or Moschatel, is half hidden among the leaves which in spring are rising up around it, but all these green blossoms are much smaller than that of the Herb Paris. This plant is singular and somewhat rare. The peculiar arrangement around the stem of its four large leaves, acquired for it the name of True Love Knot, and it is also called One Berry. The genus was termed Paris from the Latin word *par*, because of the almost unvarying number of its leaves. It has, indeed, usually a remarkable regularity in the number of its parts, the calyx consisting also of four green leaflets, and it has four yellowish green petals. Specimens of the plant, however, occur occasionally with three, five, or seven leaves; the flower-cup, too, sometimes consists of three sepals only.

The Herb Paris grows in moist woods, flowering in May and June. We have but one native species of the genus; nor is it frequent enough to be much used by " simplers " in country places, though it has an old reputation for various remedial properties. Like most plants with green blossoms, it is somewhat dangerous as a medicine in the hands of the unskilful. The leaves and berries are said to have the properties of opium, and the juice of the latter was formerly applied to the eyes in cases of inflammation. Gerarde says, that the people of Germany in his time used the leaves with great success, as an application to wounds, and that a preparation of the plant was supposed to be a " cure for such as were poisoned."

The Herb Paris is found in many parts of Scotland, and at Killarney in Ireland. Its stem is usually about a foot high. The French call this plant *Parisette*, but on some parts of the Continent it seems, from its familiar name, to have some supposed connexion with the wolf.

SPURGE LAUREL.

SPURGE LAUREL.—*Daphne Laureola.*

Class OCTANDRIA. *Order* TRIGYNIA. *Nat. Ord.* THYMELEÆ.
DAPHNE TRIBE.

THIS plant is conspicuous during the winter
months by its glossy evergreen leaves. These
surround the stem at the end of the branches,
which are quite leafless below, and hence the
green coronal bears some resemblance to a
Palm. The Spurge Laurel is common in
woods and hedges throughout England, but
is rare in Scotland. It grows to the height of
two or three feet, and its drooping yellowish
green flowers hang in the month of March,
from among the leaves, like waxen bells, and
diffuse, when the season is mild, a sweet,
though not a powerful odour. The berries are
of a bluish black colour, and are poisonous to
all animals save birds. The bark and roots
are also acrid, though less so than those of
the other wild species of Daphne. The Spurge
Laurel is often planted in gardens and shrubberies, as it thrives well among trees.

Daphne is the Greek name of the Laurel,
and the plant figured on our pages has more
resemblance to the Laurel than has the othe

wild species, the Mezereon (*Daphne Meze-reum*). This shrub grows also in the English woods, though more rarely; and is, during March, very beautiful with its purplish lilac flowers, which appear while the shrub is yet leafless. They are sweetly scented, and coming at a season when flowers are few, they render the shrub a favourite garden plant; but the scarlet berries which in autumn cluster around the stems, are highly poisonous. The acrid bark is in some countries used as a blister, and the still more acrimonious roots are employed as an alleviation in the tooth-ache; but they should be applied with caution, as they produce heat and even inflammation if held long in the mouth. Many parts of the plant have been used medicinally, but its properties are so violent that they should be adopted by those only who well understand the nature of remedies. A good yellow dye is prepared from the branches. This plant is the *Laureole gentille* of the French, and the *Laureola femina* of the Italians. *Madzaryoun* is, according to Richardson, its Persian name.

YELLOW GOAT'S BEARD.—*Tragopogon pratensis.*

Class SYNGENESIA. *Order* ÆQUALIS. *Nat. Ord.* COMPOSITÆ. COMPOUND FLOWERS.

So many rayed yellow flowers, more or less resembling the Dandelion, are to be found in meadows and on waysides, that the young botanist sometimes finds it difficult to determine their species. The Goat's Beard is, however, at once distinguished from the Hawkweeds, Hawkbits, and similar flowers, by its long slender leaves, tapering at the end, and curling about almost like tendrils. The leaves and stems are often the only portions of the plant to attract the eye of the wanderer, for the flower is closed by mid-day. Several of its country names refer to this peculiarity, as Noontide, and Jack-go-to-bed-at-Noon. It is also called Joseph's Flower, and Star of Jerusalem. When the blossom is over, a ball of downy seeds surmounts the stem, and is larger than that of any other of our wild rayed flowers. It consists of the numerous seeds, each crowned with a most exquisite array of small plumes like a shuttlecock. So well does this fibrous coronet resemble the form of the flower itself, that the writer was once asked, by one unacquainted with botany, if one of these was not the skeleton of the blossom.

The Goat's Beard is not uncommon in meadows and on field-borders in June and July. The flower-cup and foliage have a sea-green tint. The roots, if taken up before the stems shoot up, and boiled like asparagus, are scarcely inferior to that vegetable ; and almost equally good in flavour are the young spring shoots when cut off and cooked.

The Salsafy, or Purple Goat's Beard (*Trago-pogon porrifolius*), is sometimes found in moist meadows, and is similar in form to the yellow species. Its roots are large and fleshy, and tapering at the ends. They were formerly much cultivated for the table, and were eaten either boiled or stewed. Though not now much valued by the English gardener, this plant is still cultivated in France and Germany. The root of the Yellow Goat's Beard is quite as suitable for this purpose. Neither of the Goat's Beards open their flowers in rainy weather. The purple species blossoms in May and June.

COMMON AVENS.

COMMON AVENS.—*Geum urbanum.*

Class ICOSANDRIA. *Order* POLYGYNIA. *Nat. Ord.* ROSACEÆ. ROSE TRIBE.

THERE are few of our English lanes or fields, from whose hedge-banks we may not, in summer, gather this bright yellow flower. It blooms from July to August; its stem is one, two, or even three feet high, and the leaves are large in proportion to the blossom. When the yellow petals have withered, a small round head of awned seeds succeeds them, forming a brown ball, which is very conspicuous, and which, being clammy, as well as armed with hooked points, is well fitted for holding to any object near it, and thus dispersing the seeds. The root of this Avens has the scent of cloves, and will impart this flavour to ale, which it is said to preserve from becoming acid. Placed in wine it is considered to make a good stomachic, but we cannot venture to recommend its use, for Baron Haller says, that he has known this root, when infused in water, and given to patients in fevers, to produce many bad effects, and to cause delirium. The dry root laid among clothes will give them a pleasant scent, and will keep away moths. The roots are well known to have mildly astringent properties, and from their clove-like scent, the plant was formerly called Caryophyllata. If the Avens grows in a moist or shady spot, the root has little odour, nor does it acquire its scent if the season is unusually wet. This flower is often called in country places Herb Bennet, or Gold Star. The French term it *Benoite commune,* and the Italians give it the name of *Erba Benedetta.*

Another native species of the genus, the Water Avens (*Geum rivale*), is a much shorter and stouter plant than the Herb Bennet. It is not uncommon on moory ground, especially in Alpine situations. Its flowers appear in June and July; they are large, and droop very gracefully. Their colour is a dull purplish orange, marked with darker veins, and the flower-cup is also tinged with purple. A variety, which seems intermediate between the two species, is not uncommon, and has been found in Ayrshire with semi-double flowers. The name of Geum is derived from *geyo,* to taste.

HAWTHORN.—*Cratægus Oxyacantha.*

Class ICOSANDRIA.   *Order* MONOGYNIA.   *Nat. Ord.* ROSACEÆ.
ROSE TRIBE.

WHAT heart does not rejoice to see the buds of the Hawthorn whitening on the bough in the pleasant month of May? In the later season too, when its green leaves cast their shadows on the ground, and when young and old meet beneath its boughs on the village green, the Hawthorn-tree is still an object of interest. When standing alone, it will, after many years, attain a tolerable size; but sometimes it is dwarfed in its growth and becomes no larger, even though centuries pass over it. Wordsworth describes a tree of this kind, covered, as it often is, by lichens.

> " There is a Thorn, it looks so old,
>   In truth you'd find it hard to say
> How it could ever have been young,
>   It looks so old and grey ;
> Not higher than a two-years' child,
>   It stands erect, this aged Thorn,
> No leaves it has, no prickly points,
> It is a mass of knotted joints,
>   A wretched thing forlorn :
> It stands erect, and like a stone
> With lichens is it overgrown."

The scientific name is derived from the Greek *Cratos*, strength, in allusion to the hardness of the wood. Its rural names of May and Whitethorn, are significant enough. Our name of Hawthorn is supposed to be a corruption of the Dutch *Hoeg-dorn*, or Hedge-thorn; the Germans also call it *Hage-dorn*.

The Hawthorn is very variable in the form of its leaves, and in the colour of its flowers and fruit. It is never more beautiful than in its most frequent appearance of pearly blossoms, but it is also sometimes covered with clusters of blush-coloured, or deeper pink flowers. The haws afford abundant food for birds during the autumnal months. The wood, when it attains a good size, is valuable for its hardness; a fine yellow dye is also obtained from it, and the Hawthorn is said by agriculturists to be the best hedge plant in Europe, as it bears clipping well, and in its early stage grows quickly. Though the haws of this species are acceptable to children only, yet those which grow on the Thorns of the South of Europe are very delicious fruits.

WATER-CRESS.—*Nasturtium officinale.*

Class TETRADYNAMIA. *Order* SILIQUOSA. *Nat. Ord.* CRUCIFERÆ.
CRUCIFEROUS TRIBE.

THIS cress, so common in the streams of our own country, is, as Gardner has remarked, one of the few plants which are truly cosmopolite. A friend of the writer's, when wandering one evening in one of the woods of Ceylon, was reminded of her own early home among the English woodlands, not by the resemblance of the scene before her, but by the contrast. Gorgeous and powerfully scented blossoms grew beside her walk. Berries, like large clear blue or yellow or red beads, hung on the boughs, where strange birds and stranger insects were revelling. A little rivulet murmured at her feet, and looking down into its waters she beheld with surprise and delight large quantities of the Water-Cress, the very plant which in childhood she had gathered from the stream for her meal, and which in Ceylon was doubly welcome to the exile, from all its interesting associations with the past.

The Water-Cress blossoms from June till August. The foliage constitutes a well-known salad, and is in France dressed as spinach. Many poor people are supported by gathering and selling it. Alfred Lear Huxford thus alludes to it :—

" Where the still streamlet wanders o'er the glade,
  The pungent cresses grow. Hast thou ne'er seen
  The humble gatherer ply his simple craft ?
  With long lithe hook doth he his harvest make,
  Despoiling of its pure and wholesome crop
  The unmurmuring waters ; then with careful eye
  Doth he assort them, casting far away
  Each leaf suspicious, and in pleasant sheaves,
  All green and grateful, binding up the rest :
  Now his full basket glistens like a field
  Whereon an April day hath wept its showers."

Dr. Jacob, in his " Flora of Cornwall," has an interesting note on this plant. "Chaucer," he says, " employs the Saxon word *kers* (cress) to designate anything worthless,

    ' Of paramours ne raught he not a *kers*,'

from which perhaps is derived the phrase of not caring a *curse* for anything. Cresse," he adds, " is used in the same sense in the ' Testament of Love.' "

WILD ENGLISH CLARY, OR SAGE.
Salvia Verbenaca.

Class DIANDRIA.　　*Order* DIGYNIA.　　*Nat. Ord.* LABIATÆ.
LABIATE TRIBE.

THIS Sage, which is by no means unfrequent on the dry pastures of England, is so rare in Scotland, that the only recorded place of its growth is the neighbourhood of Edinburgh. It thrives best on chalky or gravelly soils, and is often abundant near the sea. At Dovor, a meadow just on the summit of the beach is sometimes, during May and June, quite covered with its dark purple and strongly aromatic flowers. These blossoms are not very conspicuous, for being raised but a little way above the deep flower-cup, a person little used to them would fancy the plant to be only in bud, when in fact its flower is fully blown. The stem is usually one or two feet high, and the leaves, though of a darker green than most of the Sage plants of the garden, have the same wrinkled appearance as theirs. This wild Sage is a native of all the four continents of the world.

Old writers descant largely on the benefit of this Clary in complaints of the eyes. Gerarde's mode of applying the remedy is not

a very inviting one to the sufferer. "The seed," he says, " put whole into the eye, cleanseth and purgeth it exceedingly from rednesse, inflammation, and diverse other maladies, or all that happens unto the eies, and takes away the pain and smarting thereof, especially being put into the eies one seed at a time and no more."

A less painful way of applying the Clary is probably more efficacious. The seed if put into water soon becomes invested with a thick mucilage, and a drop or two of this put under the eyelid, envelopes any particle of dust which may give pain to the eye, and brings it away. More than this we cannot promise for the Clary-seed, but from its old reputation it acquired the name of *Officinalis Christi*, and Clear-eye, from which last our word Clary is corrupted.

The name of Salvia, from *salvo*, to heal, was given to the genus from the high idea once entertained of the properties of some plants included in it, which were so prized as to become the subject of a common proverb among the ancients. The only other British species is the very rare Meadow Clary, or Sage (*Salvia pratensis*).

NARROW-LEAVED PALE FLAX.

Linum angustifolium.

Class PENTANDRIA. Order PENTAGYNIA. Nat. Ord. LINEÆ.
FLAX TRIBE.

THOSE only who have carefully observed our wild flowers, know how few blossoms among them are of pure blue tint. Dark purple, pink, and lilac flowers, in all varieties of hue, yellow of every shade, and snowy white, are common everywhere. The most rare colour among the flowers of Britain is red. The Poppy and Pimpernel are the only scarlet flowers, and the Adonis the only pure crimson one. The azure tint is not rare like this, for Blue Bells and Speedwells, and several others share it, yet the flowers of this hue are few in comparison with the many of other colours. Hans Christian Andersen describes blue and yellow as the prevailing colour of Swedish flowers; ours, perhaps, might be described as yellow.

This pretty Flax, however, is of a pale and pure blue colour, and is not uncommon during the month of July on sandy and chalky pastures, especially such as are near the sea. It is remarkably fragile, so much so that we can scarcely gather it without shaking its petals from the stalk, but like all its species, it is remarkable for the tough fibres of the stem.

This Flax is very similar to the Flax of Commerce (*Linum usitatissimum*), but the flowers are smaller and paler. The latter plant is also found wild in many corn-fields, and is commonly enumerated among our native plants, but it may be regarded as having sprung from seeds scattered from some land where it is cultivated.

The Perennial Flax (*Linum perenne*) is truly wild. It is about a foot high, with narrow leaves and flowers very similar to those of our engraving. It is common in chalky fields. This species is also cultivated for a thread or clothing plant.

The Flax is among the plants mentioned in the earliest portions of the Scriptures as cultivated by man, and was one of the products smitten by an Egyptian plague. In former days many of our own countrymen reared a sufficient supply of Flax for the use of the household, and the spinning-wheel hummed at the cottage doors. The necessary process of macerating the flax in water was very deleterious to such streams as were used for drinking, either by man or animals; and in the reigns of Henry VIII. and James I. acts were passed forbidding that any Flax should be steeped in a stream or rivulet used for drinking, denouncing heavy penalties on a violation of these laws.

The word *Linum* is the Celtic for thread, hence linen and its derivations. The seeds of all the species contain oil.

WILD MIGNONETTE.—*Reseda lutea.*

Class DODECANDRIA. *Order* TRIGYNIA. *Nat. Ord.* RESEDACEÆ.
ROCKET TRIBE.

OUR native plant has small claims to the name
of Little Darling, which was given by the
French to the garden species (*Reseda odorata*)
from its sweet fragrance. But though the
wild kind is scentless, it is not wanting in
beauty, and is truly ornamental to the waste
place, the chalky bank, or sea-cliff, or old
wall, where it grows so frequently. It differs
from the garden species in having more of the
greenish-yellow hue, and its stamens are not,
like the scented kind, tipped with red. Its
leaves are very variable in size, and the stems
are about a foot high. This plant is in flower
during July and August, and is known in
some counties by the name of Base Rocket.

Another species of Reseda is also common
on the waste places of chalky or limestone
soils. This is the Dyer's Rocket, Yellow-
Weed, or Weld (*Reseda luteola*). It may
easily be distinguished from the Wild Mignonette by its much longer and more slender
spikes of flowers, as well as by its less bushy
habit; and it is often three feet in height,
while it has not, like that flower, a general

resemblance to the fragrant species. As its name imports, it is used by dyers, and affords a good yellow colour, which when mixed with indigo becomes of a fine green. It is gathered for this purpose during June and July, when it is in full blossom, and no part of the plant is rejected. After boiling, cotton, wool, mohair, silks, and linen, are tinctured with it, and the yellow colour of the paint called Dutch Pink is also obtained from it. Mr. Swayne says, that this is one of the first plants which grow on the rubbish thrown out of coalpits. The root and lower leaves rise from the seeds which fall before winter, and thus this flower, like several others, is biennial; the garden Mignonette becomes a perennial if protected during winter.

The scientific name of the genus is from *Resedo,* to calm or appease; the ancients having believed that some of the species possessed sedative powers. They applied the Mignonette externally to bruises. A third species of the genus is enumerated by some botanists as a British flower, but it does not appear to be truly wild. This is the Shrubby Rocket (*Reseda fruticulosa*).

LILY OF THE VALLEY.

LILY OF THE VALLEY.—*Convallaria majalis.*

Class HEXANDRIA.   *Order* MONOGYNIA.   *Nat. Ord.* LILIACEÆ.
LILY TRIBE.

THE sweet Lily of the Valley is a flower of
the woods, and from its season of flowering
was formerly called the May Lily, an allusion
still retained by the Germans, who term it
May-blume.   Few of our wild flowers are
lovelier, and though it is not common in all
our woods, yet in some of these its broad
leaves cover a large extent of ground.   Clare
has fancifully described it among the woodland
blossoms of the early season.

> " The blue-bells too that thickly bloom
> Where man was never known to come ;
> And stooping Lilies of the Valley,
> That love with shades and dews to dally,
> And bending droop on slender threads
> With broad hood-leaves above their heads,
> Like white robed maids in summer hours,
> Beneath umbrellas shunning showers."

The leaves of this flower spring from the
root, the blossoms are sweetly scented, and the
roots are creeping.   Berries succeed the bells,
and are in autumn of a rich red colour, and
as large as a small cherry : they are not often

to be seen on the plants in their wild state, but
are frequent on those which grow in the
garden.

The scent of the Lily of the Valley is said
to have a narcotic influence. In Germany a
wine is made of the flowers, which is considered
very invigorating; and Gerarde says, that a
decoction of the blossoms is " good against
the gout, and comforteth the heart." They
readily impart their scent, as well as some
degree of bitter flavour, to water and spirituous
liquors. An extract prepared from the roots is
as bitter as that of aloes. The famous distilled
water called *Aqua aurea*, which was made of
the blossoms, was considered of great value as
a preservative from contagious maladies. A
beautiful green dye is prepared from the leaves.

As our graceful Lily is not a native of
Palestine, it cannot be the flower alluded to in
the Scriptures. The blossom to which our
Saviour directed the attention of the disciples,
appears to be either the Yellow Amaryllis, or a
species of Martagon or Turk's Cap, both of
these growing among the grass of the fields of
the Holy Land.

but

the

y is said

ermany a

considered

ays, that a

ood against

eart." They

well as some

and spirituous

om the roots is

famous distilled

h was made of

f great value as

us maladies. A

l from the leaves.

not a native of

ower alluded to in

som to which our

on of the disciples,

llow Amaryllis, or a

rk's Cap, both of

grass of the fields of

HEART'S EASE.

PANSY, OR HEARTSEASE.—*Viola tricolor.*

Class PENTANDRIA. *Order* MONOGYNIA. *Nat. Ord.* VIOLACEÆ.
VIOLET TRIBE.

THIS little flower, so common in summer time in cultivated fields, is, as well as the only other native Pansy, classed by botanists among the violets. It varies much in colour, being sometimes of a delicate cream tint, at others almost white, or tinged more or less with blue or purple; but it is so different in general appearance from the flowers which the unscientific would term violets, as to be distinguished from them without difficulty. The garden Pansies, in all their varieties, from

" The shining Pansy trimm'd with golden lace,"

to the deep dark purple, or brown blossom, are natives of other lands, and are, like our own species, scentless. The name of the flower is a corruption of the French *pensée* (*thought*), and our old writers spell it in various ways. Ben Jonson writes it the Paunsé. Milton, in his Comus, gives us another old orthography:—

" The shepherds at their festivals
    Carol her good deeds loud in rustic lays,
    And throw sweet garland wreaths into the stream,
    Of Pancies, Pinks, and gaudy Daffodils."

Michael Drayton says :—

> " The Pansie and the Marigold,
>   Are Phœbus' paramours."

And Edmund Spenser:—

> " Strew me the ground with Daffe-down-dillies,
>   And Cowslips, and King-cups, and loved Lilies,
>   The pretty Paunce
>   And the Chevisaunce
>   Shall watch with the fayre Flour de Luce."

The Pansy is also referred to by old writers under various other names, as Heart's-ease, Herb Trinity, and Three Faces under a Hood.

We have another species of wild Pansy, called also the Yellow Mountain Violet (*Viola lutea*), the flowers of which, though generally of a pale yellow, are in some specimens of a rich purple colour. This plant grows on mountainous pastures, and, though rare in the south of England, is by no means unfrequent in the northern counties, or in Wales or Scotland. A small variety of this species has also been found in Cornwall.

COMMON TORMENTIL.—*Tormentilla officinalis.*

Class ICOSANDRIA.   *Order* MONOGYNIA.   *Nat. Ord.* ROSACEÆ. ROSE TRIBE.

THOSE who study plants find it difficult to discriminate exactly between this genus and the allied one of Potentilla. In its ordinary form it is, as we have represented it in our illustration, a flower of but four petals, but it occasionally varies to five. The blossom, however, is smaller than that of the corresponding Potentilla. It is very pretty, growing on slender stems, and standing up among the grass on moors and heathy places, gleaming there in golden hue, among the Harebells and Euphrasy of Midsummer.

The roots of the Tormentil are very large and woody, and being astringent in their nature, are used medicinally. In Lapland they are of much value as furnishing a red dye for leather, and the poor people of that land tan their harness, girdles, and gloves, by a very simple and primitive method. They chew the roots of the Tormentil, together with some portion of the inner bark of the

Alder, and the saliva tinged with these vegetable dyes is applied to the leather.   In the Western Isles of Scotland these roots are thought to yield a superior material for tanning to that of the oak bark, and they are there boiled in water, in which the leather is afterwards well steeped; so that small as is our plant, it is in other lands of considerable economical importance.   The Tormentil grows in abundance on the mountains of Killarney, and large numbers of swine are there fed on the roots.   Our English farmers are glad to see it on the pastures, as it is considered very nutritious for sheep.   In former days the roots were valued as a medicine in fever.

This little flower is sometimes found creeping with prostrated stems over the ground, and when in this condition, it is by some botanists called the Trailing Tormentil, but it is in all probability the same species.

COMMON FRITILLARY.

COMMON FRITILLARY.—*Fritillaria Meleagris.*

Class HEXANDRIA. *Order* MONOGYNIA. *Nat. Ord.* LILIACEÆ.
LILY TRIBE.

THE solitary drooping flower of the Fritillary is not a very general ornament to the English meadow, and in the northern counties it is very rare. The specimen from which our engraving was made was gathered from a pasture land near Cobham, in Kent, and it is chiefly in the southern and eastern counties of England that the Fritillary is found. The dark tulip-shaped bell is curiously chequered with pink and dull purple, and hence the genus is said to have derived its name from Fritillus, the dice-box, which is the frequent companion of a chequered board. Sir William Hooker, however, derives it from the *Numidia Meleagris,* or Pintado, whose plumage is chequered in a similar manner. Of old times it was called the Guinea-hen, or Turkey-hen flower, and it is now in some country places known as the Chequered Daffodil, or Snake's Head.

The French term it *La Fritillaire méléagre*, and it is the *Das Kiebitzey* of the Germans.

This Fritillary is a bulbous plant, and blooms in April. It is the only British species, and is often cultivated in gardens, where several others of the kind are well-known flowers. Such is the handsome Crown Imperial (*Fritillaria Imperialis*), which is a native of Persia. The Fritillary is included in the Tulip group of the Liliaceous tribe, and a glance at the flower would at once convince us of its similarity to the Tulip. We have few British species in this group, but our garden is indebted to it for some of its greatest ornaments. But we need not dwell on the beauties of flowers so universally known, and whose varieties are so numerous, that, as old Gerarde remarked long since, " All which to describe particularly, were to roll Sisyphus' stone, or to number the sands."

CARLINE THISTLE.—*Carlina vulgaris.*

Class SYNGENESIA. *Order* ÆQUALIS. *Nat. Ord.* COMPOSITÆ.
COMPOUND FLOWERS.

THIS singular Thistle is common throughout
Europe on dry sandy or chalky places, and is
a sure indication of a barren soil. The inner
scales of its flower-cup are of a pale straw
colour, and so glossy and firm is their texture,
that the thistle resembles some of the flowers
of the garden which we term Everlastings.
Within these is a circle of purple florets, and
if the plant remain ungathered till winter, when
all has faded save the outer ray of its flower
cup, this becomes so white and shining that
it looks like a star cut out of silver. It sometimes happens, however, that the yellow ray
remains perfect also, not only through the
winter, but during part of the following year,
the purple florets having yielded to the yellow
down which surmounts its seeds. This Thistle
therefore is a favourite mantel ornament in
country places.

The flower-scales of the Carline Thistle are
very sensitive to atmospheric changes, and in
wet weather the ray closes up in a conical

form. In Germany, France, and Spain, it is a common practice to hang up another species, the Stemless Carline Thistle, at the doors of the cottage, to give warning of changes in the weather.

The common species blossoms in June, and is biennial. It is the only British kind, and was formerly valued as a medicine in hysterical cases. The name of the genus is a corruption of that of Carolina, which was bestowed on the Stemless species in honour of Charlemagne, to whom, as ancient legends tell, an angel pointed out the root as a remedy for the plague. This Carline Thistle has black woody roots, often an inch in thickness. The upper parts of these, as well as the flower-cup, are tender when young, and may be eaten, but when the plant is fully grown the roots become resinous and acrimonious. Though modern physicians would not trust to their properties as a cure for the plague, yet tradition tells also, that the army of Charles V. when in Barbary, were rescued from this dreadful malady by their means. The French term our common species *La Carline*.

TUFTED VETCH.—*Vicia Cracca.*

Class DIADELPHIA. *Order* DECANDRIA. *Nat. Ord.* LEGUMINOSÆ.
PEA AND BEAN TRIBE.

WE have many flowers which add a grace
to the summer hedge by creeping among the
boughs of shrubs or trees. There they may
be seen holding themselves up, either by
clasping tendrils, or by winding stems, which
curl and cling to the branch of the firmer
plant, hanging down clusters of blossoms, or
crowning their summits with floral beauty.
Most of our Vetches have this twining habit,
and, like the graceful species represented by
our engraving, they can scarcely be dissevered
from the hedge without rending their delicate
stems. Not one of them is more beautiful
than this kind, with its flower of rich purplish
blue, and its long silky leaves, composed of
about ten pairs of leaflets. The tangling
stalks and curling tendrils enable this Vetch
to climb to the very topmost twig of the
hedges, and its stem is sometimes three or four
feet long. Its crowded clusters of blossoms
are very common during July and August.

Several of our wild Vetches are exceedingly
nutritive to cattle. Gerarde thought very

highly of the qualities of this species, and Dr. Plot, in his " Natural History of Staffordshire," says that this, and the less frequent Wood Vetch (*Vicia Sylvatica*), "advance starven or weak cattle above anything yet known." All the ten species of wild Vetch are more or less useful for this purpose, while their seeds afford a large supply of food to the birds. The earliest blooming of the species, the Bush Vetch (*Vicia Sepium*), is a valuable plant, because it shoots sooner in the spring than almost any other kind of food eaten by cattle, while it vegetates too in autumn, and has some green leaves and shoots rising amid the general decay of winter foliage. It is palatable to every kind of cattle, but though abundant as a wild plant, it is difficult of cultivation on a large scale, as it is often devoured by the larvæ of an insect. The Common Vetch (*Vicia sativa*), with its blue and purple or red flowers, is extensively cultivated in our fields as fodder.

The name of *Vicia* is said to be originally from the Celtic *gwig;* in German *wicken;* *bikion* in Greek, and *vesce* in French.

SELF-HEAL.—*Prunella vulgaris.*

Class DIDYNAMIA.   *Order* GYMNOSPERMIA.   *Nat. Ord.* LABIATÆ.
LABIATE TRIBE.

THE Self-heal is a common plant on our pasture-lands in the months of July and August. At this season, though the grass has not the bright green of the spring-time, yet there is a richness in the tints of the blossoms and of the trees, which renders the country scene truly beautiful.

> " The green herbs
> Stir in the summer's breath; a thousand flowers
> By the road-side and the borders of the brook,
> Nod gaily to each other; glossy leaves
> Are twinkling in the sun, as if the dew
> Were on them yet."

The Labiate tribe, to which our Self-heal belongs, are now most numerous. We meet with some of them at every step of our rural walk. Most of them have some old reputation for medicinal properties, and our fathers gave to this plant several familiar names, expressive of its virtues. They called it Carpenter's herb, Hook-heal, and Sicklewort; and in old times it was a frequent application to the wounds inflicted by the implements of rustic labour. Like most of the remedies taken from the fields and used by our fathers, its worth was over estimated, but as its pro-perties are astringent, it was doubtless of use in such cases.

. The name of *Prunella* is taken from the German word for quinsey, for which our plant was formerly considered a certain cure, and which appears to have been a very prevalent disorder from the number of plants used as remedies for it. Country people call it *Brunella*, the French term it *La Prunelle*, and it is *Die Prunelle* of the Germans. It is the only British species of the genus. The heads of flowers, though usually bluish purple, sometimes vary to pale lilac or even white; and the bracts which grow among the blossoms are generally much tinged with purple.

The *Prunella* is common to pasture-lands throughout Europe, and we may find it some-times blooming on our sheltered hedge banks, as late as the end of October.

COMMON COW WHEAT.—*Melampyrum pratense.*

Class DIDYNAMIA. *Order* ANGIOSPERMIA. *Nat. Ord.* SCROPHU-
LARINEÆ.

FIGWORT TRIBE.

THIS plant is not, as one would suppose from its name, a native of woods and meadows. It is not uncommon among the trees and bushes of our woods and thickets, blooming in July and August. Its flowers are of pale sulphur colour, and the name *Melampyrum*, Black Wheat, refers to the seeds, which are very similar in form to grains of wheat, and which when ground with flour, are said to make it black. It was formerly cultivated by the Dutch and Flemish as food for cattle; and its familiar names throughout Europe mostly refer, like our English one, to its being eaten by cows. The Germans term it *Der Wachtelweizan;* the Spaniards, *Trigo de Vaca.* Among the French and Italians it is known by the name of *Melampiro.* It is usually about a foot high, with slender straggling branches.

There are four British species of Cow Wheat, and the purple species (*Melampyrum arvense*)

is a very handsome flower. It grows in corn-
fields and on dry gravelly banks, but is not
common except in Norfolk, where, especially
about Norwich, its large handsome spikes of
flowers, of various shades of yellow, purple,
and pink, are very conspicuous. It is not,
however, a nutritious plant, and its abundance
in the corn-field greatly deteriorates the crop.
Some lands in the Isle of Wight have been
much lessened in value by its growth there.
It is supposed to have been brought among
the grain from Norfolk into the Island, and it
is there called Poverty Weed. It grows only
on such corn lands as have a chalky soil. It
is eaten by cows, but rejected by sheep.

The Crested Cow Wheat (*Melampyrum
cristatum*) is also found in Norfolk, as well as
in Cambridgeshire, Bedfordshire, and Hun-
tingdonshire; and is a beautiful plant, flowering
in July, its yellow blossom having a tinge of
purple on its upper lip. The lesser flowered
Yellow Cow Wheat is a rare plant of alpine
woods in the north of England.

PHEASANT'S EYE.—*Adonis autumnalis.*

Class POLYANDRIA.   *Order* POLYGYNIA.   *Nat. Ord.* RANUNCULACEÆ.
CROWFOOT TRIBE.

No other of our wild flowers can show so
rich a crimson tint as this.   The plant, however, is not truly indigenous, and was in all
probability introduced here from other lands
among grain.   In some corn-fields it is not
an infrequent weed.   The author has gathered
it from such spots near Maidstone, and
Gerarde says of it, " The red flower of Adonis
groweth wilde in the West parts of Englande,
among their corne as Maie-weed does; from
thence I brought the seede, and have sowen it
in my garden for the beautie of the flower's
sake."   It seems in all succeeding times to
have been prized by those who have gardens,
for we find it yet as a common border flower,
its rich crimson buttercup-shaped blossoms
gleaming in July from among its narrow leaves.
Miller, in his " Gardener's Dictionary," states
that great quantities of these flowers were
brought every year to London, and sold in the
streets by the name of Red Morocco.   Gerarde

calls it the *Rose a rubie*, which he says was its name among the herbe women.

The name of Adonis was given to the genus from the youth Adonis, who, as ancient poets tell, was slain by a wild boar, and whose blood crimsoned its petals. The fable of this youth, like that of Narcissus, was a favourite legend; for there is in Lebanon a river, running over a red soil, which is called Adonis, and said, too, to have been stained by his blood. A reference to the tradition as affecting the flower, is found in several European names. Our gardeners term it *Flos Adonis;* the French call it *L'Adonide;* and the Germans, *Die Adonis blume*, or, *Adonis rose.* It is the *Adonis bloem* of the Dutch, the *Fiore d'Adono* of the Italians, and the *Adonis* of the Spanish and Portuguese. Its familiar name in France is *Goutte de Sang.*

This is the only native species.

COMMON CALAMINT.—*Calamintha officinalis.*

Class DIDYNAMIA.   *Order* GYMNOSPERMIA.   *Nat. Ord.* LABIATÆ.
LABIATE TRIBE.

THE aromatic odour of this plant at once connects it with the similar fragrance of the Mint tribe, and the name of the genus, derived from the Greek, signifies good mint. A herb tea is very frequently made of it in villages, for although the use of Chinese tea has generally superseded that of these older infusions, yet there are many persons who consider the tea made from our wild plants as peculiarly conducive to health. Nor are they, it seems, mistaken in their conclusions, for we have the recent testimony of a physician in their favour. " Many robust men and women among our peasantry," says Dr. George Moore, " from notions of their own, use infusions of Balm, Sage, or even a little Rue and Wild Thyme, as a common drink, with satisfaction to their stomachs and advantage to their health." The author knew a person residing in the country who could never be persuaded to drink any tea save that made of his native herbs, and

who ascribed his uninterrupted health at an
advanced age to this habit. The Calamint is
a favourite plant with such persons.

Our plant is very well known, for it grows
commonly on way-sides and field borders,
especially when the soil is of gravel. Its stem
is erect and bushy, half a foot or more in
height, and its pinkish lilac flowers bloom in
July and August.

Another species, called the Lesser Calamint
(*Calamintha Nepeta*), is very common on chalky
soils. It has the same kind of odour, resembling that of Penny Royal. It is very similar
to the plant represented by our engraving,
though the serratures on its leaves are more
strongly marked, and it has some prominent
white hairs at the mouth of the flower-cup.
Many botanists consider that it is but a variety
of the common species, and that the marks,
mentioned as characteristics of the Lesser
kind, are not permanent, or to be found in all
specimens.

SPOTTED PERSICARIA.—*Polygonum persicaria.*

Class OCTANDRIA. *Order* TRIGYNIA. *Nat. Ord.* POLYGONEÆ. PERSICARIA TRIBE.

WE have many kinds of *Polygonum* growing wild, some botanists enumerating eight, others ten or eleven of the genus, in the British flora. The Spotted Persicaria may be distinguished from the others by its long slender leaves, which have a purple spot in the centre. This mark is the subject of a Highland tradition, and is said to have originated in a drop of blood which fell on a plant growing at the foot of the cross, in the solemn hour of the Crucifixion. This species is very common on moist ground and waste places, its greenish and rose-coloured spikes of flowers blooming in the months of August and September. All the plants of this genus have joints in their stems, hence their name *Polygonum*, which, in Greek, signifies many knees or joints.

There is much general resemblance in the appearance of most of these plants, and several, like the Spotted Persicaria, grow on moist lands. Such is the case with the Amphibious

Persicaria (*Polygonum amphibium*), which from July to September raises its handsome spikes of rose-coloured flowers above the waters, or grows on their margins. Lands recovered from rivers or lakes, or marshes which have been drained, are soon quite overrun with this plant, and if such an alluvial field be sown with grass, or left untouched, this Persicaria will require much skill and trouble in its eradication. The Water Pepper (*Polygonum hydropiper*) is another species found in abundance where water has stood. It is a less showy plant, its greenish flowers being only slightly tinged with red. The fresh juice of this Pepper, though not disagreeable, is very pungent, from an essential oil which exists in numerous dotted glands scattered over the surface of the plant. The Snakeweed or Bistort (*Polygonum bistorta*), which owes its name of twice-twisted to its curling root, is also found in the moist meadow land, and has handsome flesh-coloured flowers, but it is not a common plant. The remaining species are found in fields and meadows of dryer soil.

PROCUMBENT PEARL-WORT.—*Sagina procumbens.*

Class TETRANDRIA. *Order* TETRAGYNIA. *Nat. Ord.* CARYOPHYLLEÆ.
CHICKWEED TRIBE.

LITTLE tufts of this plant may often be seen crowding over the crevices of the wall among the moss and ferns and other plants which cluster there. Its blossoms have a delicate beauty, though they need close observation to detect it, and the prostrate stems have slender, pointed leaves rising out of them. The plant is in bloom throughout the summer, casting out numerous little seeds from the small seed-vessels which succeed the blossom, and being thus prolific and perennial too, it is very abundant. The gardener finds it a troublesome weed on his pathways. The pavement of the town which may be little trodden, and the pasture land, the wayside, and other waste places, are its frequent haunts. Its stems are from one to four inches long, and it is known to country people by the names of Seal-wort, Break-stone, and Little Chickweed.

It is not easy for any but a botanist to distinguish between this and the two other wild species. The annual Small-flowered Pearl-wort (*Sagina apetala*) is rather less in size, and is besides an annual plant, flowering in May and June, and forming tufts on dry walls, gravelly and sandy heaths, and similar spots. Its stems are not so prostrate as those of the procumbent kind.

The Sea Pearl-wort (*Sagina maritima*) has its distinctive c arac r rather more marked. It grows upright, andit has usually a purplish or reddish hue on both stems and calyxes, but it has no petals. It is in flower from May to August in sea-side places, and is not infrequent on the coasts either of England, Ireland, or Scotland. The French call the Pearl-Wort *La Sagine*, and the Germans, *Der Vierling*. Linnæus says, that its name of *Sagina*, expressive of something nourishing, is indicative of its qualities, as it is good food for sheep. Curtis remarks of the small-flowered species, that there is scarcely any plant which ripens its seeds so rapidly.

COMMON YELLOW RATTLE.—*Rhinanthus Crista-Galli.*

Class DIDYNAMIA. *Order* ANGIOSPERMIA. *Nat. Ord.* SCROPHU-
LARINEÆ.

FIGWORT TRIBE.

THE seeds of this plant, when ripened and lying loose in their large capsules, rattle as the wind blows over and shakes them, and thus originated its common names. In Ireland it is frequently termed Rattle Box, in some country places it is called Penny Grass, and in Yorkshire, Hen Penny, from the seed vessel, which is somewhat of the shape and size of the smallest silver coin. It is also termed Cock's Comb, and the French name it *Creste du Coq.*

This Yellow Rattle is a firm erect plant, from ten to twelve inches in height, and is rendered more conspicuous by its inflated flower cups than by its small yellow blossoms. It is common, and very abundant on some grass pastures and cultivated lands, yet is not so general as to be universally known. It flowers with us in June, and ripens its capsules by the end of July; but as the ripening of its seed is regarded by the Swedes as the season at which their hay should be cut down, it possibly flowers at a different time in that country. It is not a plant which is welcomed by the owner of the pasture land, for cattle are not fond of it, and appear never to touch it but from necessity. It is more plentiful in some summers than in others.

A kind, which perhaps should rather be termed a variety than a species, is called by some botanists the Large Bushy Yellow Rattle (*Rhinanthus major*). It is found in corn-fields in the north of England, and has dense spikes of flowers, and a slight tinge of purple on the upper lip of its yellow corolla. Dr. Richardson has remarked of this, that when the soil approaches to peat, it almost obliterates the crops. It blossoms two or three months later in the year than the more common plant.

TUBEROUS BITTER VETCH

TUBEROUS BITTER VETCH.—*Orobus tuberosus.*

Class DIADELPHIA. *Order* DECANDRIA. *Nat. Ord.* LEGUMINOSÆ. PEA AND BEAN TRIBE.

THIS vetch, with its clusters of blue and purple flowers, is not an infrequent plant of the woods, blooming in May with the Wood Anemone, and remaining in blossom a month later than that flower. It is often called Peaseling, or, Wood Pea. It is little noticed in England, and our country people leave its roots untouched; but they are much prized by the Highlanders, and Lightfoot has recorded their various uses. The people of the Scottish Highlands know well how to make the most of the scanty vegetation of their native soil. They call the root of this plant Cormeille, and chew it in order to give a better relish to their liquor, while they believe that a very small portion will suffice to allay both hunger and thirst. The roots are also considered to be a valuable remedy against disorders of the lungs. Their flavour is sweet, resembling that of the root of the Rest Harrow; and they are said to be very good when boiled, or roasted as

chestnuts, in which way they are commonly
eaten in Holland and Flanders. Lightfoot
says, that in Breadalbane and Ross-shire, it
is a common practice, after bruising and
steeping them in water, to brew from them a
pleasant fermented liquor. Dickson recommended their culture in England.

Besides its uses in modern times, the plant
has an interesting association with the past.
Baxter remarks of it, " It is supposed to be
the Chara named in Cæsar's Commentaries,
the root of which, steeped in milk, was so
great a relief to the famished army at the
siege of Dyrrachium. It is also believed to
be the Caledonian food described by Dio, on
which, mixed with milk, the soldiers of
Valerius' army subsisted, under penury of
bread." We have two other wild species of
Orobus. The Black Bitter Vetch (*Orobus
niger*) grows in Scotland on shady rocks; and
the Wood Bitter Vetch is a native of woods
and thickets in the north of England and
Wales, and in the Lowlands of Scotland.

BLACK HOREHOUND.—*Ballota nigra.*

Class DIDYNAMIA. *Order* GYMNOSPERMIA. *Nat. Ord.* LABIATÆ.
LABIATE TRIBE.

EVEN those who are not very observant of flowers, will at once recognise this as a common plant in all parts of our land, save the northern counties. There it is less frequent, but in most places it grows on every waste ground, and rises up by every public wayside, its dim purple flowers and greyish green leaves looking duller from the dust with which they are covered. It is three or four feet high, flowering from July till September.

Although this weed may sometimes be found far from man and his habitation, yet it is their general accompaniment; and we are scarcely off the town pavement and in the road towards the country, before we see it growing in the hedge. This tendency of certain plants to follow man is very remarkable, and in some others it is more striking than in this, for several, as the Soapwort and the Vervain, are rarely found except near villages or towns.

Sir T. L. Mitchell, in his work on "Tropical Australia," while recording that he halted near a spot where sheep were reared, says, "Here I perceived that Horehound grew abundantly, and was assured by Mr. Parkinson, a gentleman in charge of these stations, that this plant springs up at all sheep and cattle stations throughout the colony; a remarkable fact, which may assist to explain another, viz. the appearance of the Couch Grass, or the Dog's-Tooth Grass, wherever the white man sets his foot, although previously unknown in these regions."

The Horehound has a peculiarly disagreeable odour, hence its name, from the Greek verb, to reject. It is not much valued as a pasture plant in this kingdom, but the Swedes consider it as an almost universal remedy in the diseases of cattle. It was formerly in some repute in this country for medicinal properties, and it is still recommended by modern practitioners. It has sometimes white flowers.

YELLOW HORNED POPPY.—*Glaucium luteum.*

Class POLYANDRIA. *Order* MONOGYNIA. *Nat. Ord.* PAPAVERACEÆ. POPPY TRIBE.

IT is delightful when rambling over the stony beach, listening to the tune of the waters, to find that a flower is blooming even there. The sands and the shingle are not very prolific of vegetation, though the cliff which bounds them is often green and gay with wild flowers; and if it be of chalk, the blossoms are peculiarly bright in hue. The Sea-side Convolvulus and the Sandwort, with some few other maritime plants, grow near to the margin of the sea, but no flower is gayer or more frequent there than the Yellow Horned Poppy. Like others of the Poppy tribe, it has crumpled and fragile petals, but so many are the blossoms, that when the wind carries away some, others soon succeed, and adorn the plant with golden beauty for many successive weeks. The foliage has that delicate sea-green tint which the botanist terms *glaucous,* and from which it derives its scientific name. It is rough with short bristles, and few objects are more beautiful than a large leaf of this plant, glittering at every point with hoar-frost, for it is green even in winter. The cylindrical pods are sometimes ten inches long, looking like horns; and the root is tapering and yellow like that of the carrot. The plant is biennial.

The sea-side Poppy blooms from June till August, on most beaches, as well as on the sands and neighbouring cliffs. It is the only truly wild species, though two other kinds of Horned Poppy have been occasionally found in England, and are hence commonly enumerated in the British Flora. These are the Scarlet Horned Poppy (*Meconopsis phœniceum*), and the Violet Horned Poppy (*Meconopsis violaceum*).

Gerarde mentions, that in his time the flowers, leaves, and roots of our common species were much used as medicine; but the plant is highly acrid and dangerous.

COMMON FROG-BIT.—*Hydrocharis Morsus-Ranæ.*

Class Diœcia. *Order* Monadelphia. *Nat. Ord.* Hydrocharidaceæ. Frog-bit Tribe.

THE delicate white blossoms of this handsome plant adorn many of our English ponds and lakes, and are not uncommon on the still waters of the Irish landscape, though unknown in Scotland. Like most aquatic plants, it is what botanists term social in its habit, a large number growing together, and often forming extensive patches on the waters; the substantial glossy leaves contrasting well with the fragile petals. It is in blossom during July, floating on the surface, and sending long radicles from the stem, which penetrate far down in the soil at the bottom of the pond. The stems lie horizontally on the water, extending to great length, and the joints are furnished with pendulous buds on long footstalks. These buds on examination are seen to consist of two scales, within which the embryo leaves of the future plant lie curiously enfolded. The flowers arise from pellucid membranous spathes, or sheaths.

The form of the blossom, consisting of three petals, though very rare among the plants of wood or field, is less uncommon in our aquatic species. The Water Plantain, the Water Arrow-head, and some others, are similarly shaped.

The English name of Frog-bit refers to the haunts of this plant, where the blithe reptile is its frequent companion : the Germans too term it *Der Froschbiss.* In some country places it is called Lesser Water-lily, the form of its kidney-shaped leaves resembling those of that flower. The generic name is from the Greek words for *water* and *to rejoice.*

Our Common Frog-bit is the only British species of the genus, but Ray describes a species, which had double and odoriferous blossoms, and grew, in his time, plentifully in a ditch in the Isle of Ely. It has been sought for there by modern botanists without success.

SEA SPURREY SANDWORT.—*Arenaria marina.*

Class DECANDRIA. *Order* PENTAGYNIA. *Nat. Ord.* CARYOPHYLLEÆ.
CHICKWEED TRIBE.

THE name of the genus *Arenaria*, from *Arena*, sand, is indicative of the soil on which most of the species are found. The plant figured here is frequent on the sea-coast, sometimes hanging down in tufts from the crevices of the cliffs, at others growing on the sand or marshes of the shore. The species so frequent on gravelly and sandy inland soils, called the Purple Sandwort (*Arenaria rubra*), is much like this, but is generally more slender, and has smaller blossoms. Many botanists think it to be the same kind, varied only by the influences of climate and soil. The little star-like blossoms are to be found from June till August; and when growing on salt marshes in great numbers, they are ornamental to their grassy surfaces, for the stems are there often more upright, and the flowers larger than on the sand.

Ten species of the Sandwort grow wild
in Britain, though some of them are very rare
plants. Such is the Norwegian Sandwort
(*Arenaria Norvegica*), which was discovered
in 1837 in the Shetland Isles. The pretty
Vernal Sandwort (*Arenaria verna*) is not a
common flower, though it grows in several
parts of Scotland, and at the Lizard Point,
Cornwall. At the latter place it is found on
Magnesian rocks near the sea, and it is also
very abundant on Arthur's seat, near Edinburgh.

One other, besides the Spurrey Sandwort,
grows on the shore; it is the Sea-side Sandwort (*Arenaria peploides*). We may often
see extensive patches of this plant on the
beach, or sand, without perceiving the blossoms. These are small and white, and besides
being much hidden by the crowded leaves,
they are only open during sunshine. The
plant blooms in July, and after flowering, has
comparatively large seed-vessels containing
black seeds.

COMMON WILD PARSNEP. — *Pastinaca sativa.*

Class **PENTANDRIA.** *Order* **DIGYNIA.** *Nat. Ord.* **UMBELLIFERÆ.**
UMBELLIFEROUS TRIBE.

THIS plant is very common on banks and in hedges where the soil is of chalk. Both in its small cluster of yellow flowers and in the form of its leaves, it is very similar to the cultivated Parsnep, which is indeed but a variety of the wild kind. The root has also the same odour and flavour as our well-known vegetable, but in its native state it is tough and woody. It blooms during July and August, and its yellowish green seeds remain till November on the stem. These contain a large quantity of essential oil, as any one may discover by tasting them; though their pungent and bitterish flavour remains long on the tongue, and is too strong to be agreeable. It would have been strange if our forefathers had not found some virtues in so powerful an oil, and we find it recorded as an excellent remedy in cases of intermittent fever.

The root of the Parsnep, when submitted to

culture, contains a large amount of nutriment.
It has long been used, especially by Catholics
during Lent, as an accompaniment to fish;
and in former times it was occasionally made
into bread. The name of the plant is derived
from *pastus,* nourishment. In the North of
Ireland a kind of beer has been brewed from
the Parsnep flavoured with hops; and the root
yields, by distillation, an ardent spirit, from
which a very agreeable wine is made; though
Parsnep wine, like that of the Cowslip, belongs
more to the past than the present time. In
the North of Scotland the children of the
peasantry thrive well on boiled Parsneps beaten
up with milk and butter; and the root is often
largely cultivated as food for cattle. Our wild
kind is the only British species, and is a
biennial plant. It is one of the few of our
native umbelliferous tribe which have yellow
flowers, the prevailing colour being white.

BASIL THYME.—*Acinos vulgaris.*

Class DIDYNAMIA. *Order* GYMNOSPERMIA. *Nat. Ord.* LABIATÆ.
LABIATE TRIBE.

THE Basil Thyme, with its small purple
blossom having a white centre, is to be found
during July or August on dry open places,
as chalky downs or road-sides. Just at this
season labiate flowers are abundant in our
fields and hedges, and contribute in no small
degree to the odour of the wild nosegay
gathered from the hill-side. Nor is this
species wanting in aromatic fragrance. The
Basil Thyme, though not among our commonest plants, is very frequent in some places,
and in Kent is by no means a rare flower.
Its blossoms are small, but very numerous,
and exceedingly pretty. The usual height of
its stem is six or eight inches, and the flowers,
though often of the deepest purple, are, in
some specimens, of a delicate pale lilac hue.

This is the only British species of the genus,
and it received its name from a Greek word,
which was probably applied by the ancients

to some species of Thyme. Linnæus included our plant in the Thyme genus, but its mode of growth is altogether different from their's, though the shape of the flowers is somewhat similar. The French call our flower *Basilique Sauvage.*

Their sweet fragrance has led to the cultivation of some species of Basil Thyme in the garden. These are chiefly natives of Southern Europe, and although their blossoms are larger than those of our wild plant, yet they are not showy. The true Basils are both larger plants, and possessed of more powerfully aromatic properties. Some of them were formerly much used in French cookery, and other species were valued greatly for supposed remedial virtues.

YELLOW WATER LILY

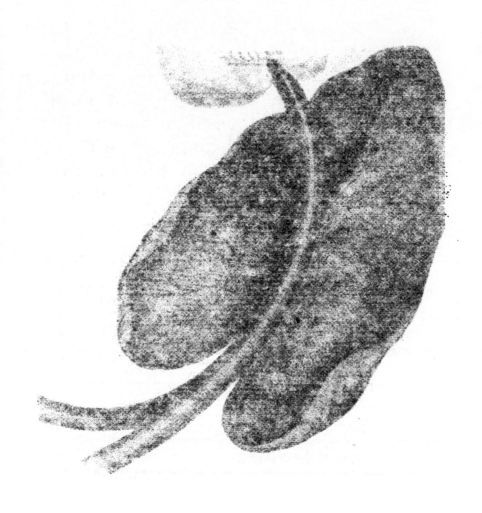

YELLOW WATER-LILY.—*Nuphar lutea.*

Class POLYANDRIA. *Order* MONOGYNIA. *Nat. Ord.* NYMPHÆACEÆ.
WATER-LILY TRIBE.

THE water-lilies are the most beautiful of all
the lovely flowers which grace our clear pools
or slow rivers. The yellow species cannot,
indeed, rival the white in this respect, but its
richly-coloured cup, attached to a stem often
a yard long, contrasts admirably with the
glossy green leaves which crowd the surface—

> " Making the current, forced awhile to stay,
> Murmur and bubble as it shoots away.".

So smooth is this foliage, that it feels to the
touch like oiled cloth, and the water runs off
from it as it would do from that substance.
The unfolded leaves are coiled into a vase-like
form, hence the flower is commonly known in
country places as the Water-can. It has also
the name of Brandy-bottle, partly, perhaps,
in reference to its bottle-shaped capsule; but
chiefly from its odour. This scent, in the open
air, is not unpleasant, but when two or three

flowers are in a room it becomes very disagreeable, and the writer, after making the drawing for the illustration, was indisposed for some hours from its effects.

This Water-lily blossoms in June, and is very frequent in rivers and ditches. It is a common flower also in the rivers of Oriental countries, and is among those which the Arabs prize as adding to the delights of wine. The Turks prepare a cooling drink from the blossoms, which they call *Pufer*, a corruption of its ancient name, *Nouphar*. The Arabs, according to Forskahl, still term it *Noufar;* and our scientific name is of Greek origin. Swine are fond both of the leaves and roots, but most animals refuse them. Country people place them with milk in their houses, to drive thence cockroaches and crickets.

In many of the Highland lakes, as well as in a pond near Wallington House in Northumberland, a smaller species is found, called the Lesser Water-lily, *Nuphar pumila*.

SMALL BUGLOSS.—*Lycopsis arvensis.*

Class PENTANDRIA.   *Order* MONOGYNIA.   *Nat. Ord.* BORAGINEÆ.
BORAGE TRIBE.

"INGENIOUS people," says Loudon, "have found a similarity between the small blue flower of this plant and the eye of a wolf." *Lycopsis,* taken from the Greek, signifying wolf's eye, or as some say, wolf's face. Nor was the notion of a similarity of this kind confined to the ancients, for the familiar name yet in use by the Dutch for the plant, is *Wolfschyn.* Yet a more innocent-looking little flower than this never grew by the wayside, and one would have thought it better fitted to remind the poet of the eye of the fair maiden, or the smiling child, than of that of the fierce wild animal. The azure tint of the blossom is very bright, and rich dark green leaves, studded with hairs or rather bristles, clasp closely at their bases the rough stem, or grow around the root on footstalks. Each hair is situated on a white callous tubercle, rendering the foliage singularly rough, even in a tribe nearly

all the plants of which are remarkable for their
hairy or bristly leaves.

The Bugloss grows in many corn-fields and
on hedge-banks, especially those near the sea,
flowering from June to August.   Its blossom
is remarkable in one particular, its tube being
bent; whereas most of the other plants of
the order have straight tubes to their corollas.

This is the only native species of the genus,
but it is very nearly allied to the Alkanet
(*Anchusa*); and was, by earlier writers, classed
with it.   The Common Alkanet (*Anchusa
officinalis*) has, however, deep purple flowers,
and the somewhat rare Evergreen Alkanet
(*Anchusa sempervirens*) may be known by its
ovate leaves, those of our Bugloss being long
and narrow.

We have some species of Lycopsis in the
garden, but the small size of the blossoms
prevents their being very ornamental, though
their blue tints are very rich.

SAMPHIRE.—*Crithmum maritimum.*

Class PENTANDRIA.   Order DIGYNIA.   Nat. Ord. UMBELLIFERÆ.
UMBELLIFEROUS TRIBE.

THE tall sea cliff has tufts of this plant scattered among its crevices, and fringing its summit, at a height where no eye can see it save that of the sea-bird. Though not common to all our rocky shores, it is plentiful on some of them, as on those of the Isle of Wight, Cornwall and Dovor. It is better suited for pickling than any other wild plant, for it has a strongly aromatic flavour, and from early times it has been collected from the rocks during May for this purpose. Shakspeare's well-known lines referring to the Samphire gatherer on Dovor Cliff, Michael Drayton's reference in the Poly-Olbion, as well as John Evelyn's praises of the plant, record its uses to our fathers. In former times, however, it was also mingled with other herbs for salad, and the last-named author recommends its use for sharpening the appetite and other invigorating effects. Though still eaten when pickled, yet its flavour with salad is not agreeable to modern palates.

Large clumps of the yellow clustered flowers
of the plant and its fleshy sea-green leaves
ornament the cliff during summer, and in
autumn and winter the foliage may yet be
seen in the sheltered nook. The French term
it *Créte Marine*, and *La Bacille;* and its
name of Samphire is evidently a corruption of
Herbe de St. Pierre.   Its botanic name is
from the Greek *Crithé*, barley, to which grain
its fruit is fancied to have some resemblance.

This herb of our sea-rocks is the only
British species; but a succulent sea-green
plant of the Salt marshes, the Jointed Glass-
wort (*Salicornia Herbacea*), is often sold under
the name of Samphire, and made into a pickle.
Hence it is commonly called Marsh Samphire
by country people. It is a more general plant
on the coast, but is far inferior in flavour to
that whose name it sometimes bears. It may
readily be known from it by its leafless and
jointed stem, with small green flowers between
the joints of the terminal branches.

YELLOW BED-STRAW.—*Galium verum.*

Class TETRANDRIA. *Order* MONOGYNIA. *Nat. Ord.* RUBIACEÆ.
MADDER TRIBE.

THE dry bank, especially near the sea, is often covered as with a yellow carpet, by the golden flowers of this Bed-straw. It is in blossom during July and August, and its slender whorled leaves are verdant when almost all other foliage is scorched, and are still so even till the close of the year. This plant has the name of Cheese Rennet, and is said by old and modern writers to curdle milk. The Highlanders add salt and nettles to form their rennet, finding the Bed-straw alone insufficient. The author has tried various experiments with this herb, and never could coagulate the milk by its means, nor produce any effect, save that of giving to the fluid a very disagreeable flavour. The foliage is slightly bitter, and a small degree of acidity may be detected in it, but, as Curtis remarks, its acid is more subtle than that of Sorrel, or the many other vegetable acids. A kind of vinegar is said to be distilled from the flowers of this species. They have a sweet scent like that of honey, and Loesel says, that this fragrance is stronger than usual during wet and stormy weather.

The roots of this Yellow Bed-straw mixed with alum, are used by the Highlanders to dye red; and it was once cultivated for this purpose, at the recommendation of the Committee of the Council for Trade. The colour is said not to be inferior to that of the Madder, but the plant is much smaller. Another British species, the Cross-leaved Bed-straw (*Galium boreale*), is described by Sir John Franklin in his work on the " Polar Seas," as being used by the North American Indians as a vegetable dye. They term it *Sawoyan*, and after boiling the roots, they mix the liquid with the juice of the strawberry and cranberry and a few tufts of the pistils of the larch, and with this, dye the porcupine quills of a beautiful scarlet.

Botanists enumerate sixteen British species of this genus. They have chiefly white flowers. The word *galium*, from milk, was given from the old uses.

FIELD KNAUTIA.—*Knautia arvensis.*

Class TETRANDRIA. *Order* MONOGYNIA. *Nat. Ord.* DIPSACEÆ. TEASEL TRIBE.

THE Knautia is one of the handsome flowers which, in so great number, grow among the corn, and it is also found on hedge-banks and borders of meadows. The agriculturist regards it as a troublesome weed in his corn-field, but welcomes it to the pasture land, where its leaves afford a good supply for the animals feeding there. Its large lilac heads of many flowers, grow on a stem two or three feet high, and may be seen from afar, over-topping the brown corn in July and August, or falling with it before the reaper's sickle. The leaves are dark green, slightly hairy; those around the root being long and slender, and only slightly notched, but the upper leaves are more or less divided. They have a rather bitter taste, and some astringent properties, and have been used as a medicine for coughs and asthmas. The flower is often called Blue-caps in country places, and is also by many termed Scabious, but although much like it, yet it differs in various respects from the true Scabious, which will be described on a future page. In the South of England both this and the Premorse Scabious are commonly called the Gipsy Rose.

This is the only British species of the genus Knautia, but some handsome flowers belonging to it are cultivated in gardens, having been introduced hither from the countries bordering the Mediterranean, where some of them grow wild in the fields and meadows. The genus received its name from Linnæus, who gave it in remembrance of Christopher Knaut, a physician of Halle, in Saxony, who died in 1694.

GREAT BINDWEED.—*Convolvulus Sepium.*

Class PENTANDRIA. *Order* MONOGYNIA. *Nat. Ord.* CONVOLVULACEÆ. CONVOLVULUS TRIBE.

THOUGH our large Bindweed has not the gift of fragrance, yet few flowers are more elegant in form, or purer in their tint of snow. Often, as autumn is approaching, the large white flowers of this plant hang like marble vases amid the brown and yellow leaves of the hedge, left flowerless now, save where a pale bramble blossom lingers on the bough, or a golden dandelion gleams from the grass beneath. We could almost wish, when we see it there, that our hedges were like those of Italy, where thousands of flowers of this tribe in various hues peep from among the leaves of the bushes. Our large Bindweed does not, like the smaller rose-coloured species, close before rain or in the evening dew ; and we have often, by moonlight, seen its white bells filled with the pearly dews of night. Yet like all its tribe, its flower is fragile. In Sierra Leone, where the Convolvuluses bloom all the year round, though each blossom lives only for one day, yet they are kept fresh and vigorous amid the general withering of every leaf and blade, by the dew which lingers in the large leaf out of which they spring. In that land they are of various colours, but chiefly white.

Our white Bindweed has been called English scammony, and its juices are well known to be scarcely inferior to those of which this drug is formed. Like many other climbing plants, our flower shows a remarkable instinct of vegetation. If a prop be placed within six inches of one of its young shoots, it will reach it, although the prop be removed daily to a little distance. If after having twined some way up the pole, it be unwound and twisted in an opposite direction, it will either return to its former position, or perish in the attempt. Yet if two of these plants be placed near each other, and have no prop around which to entwine, one of them will alter the direction of its spiral, and they will wind round each other.

COMMON GROMWELL.—*Lithospermum officinale.*

Class PENTANDRIA. *Order* MONOGYNIA. *Nat. Ord.* BORAGINEÆ. BORAGE TRIBE.

THIS flower was called in France, some centuries since, *Plante aux Perles,* from its hard stony seeds. After the blossom has faded, we find in the calyx the little seeds, first green and soft, but maturing into hardness, and becoming of a pale brown colour. They are highly polished when ripe, but seldom more than one or two in each flower-cup becomes perfect. Professor Hooker informs us, that his friend Captain Le Hunte submitted these seeds to analysis, and obtained the following results. The stony shells of sixty seeds weighed upwards of seven grains. Heated to redness, these seven grains were reduced to three, of which four-tenths of a grain were pure silica. There was also a considerable quantity of phosphate of lime and iron. The flinty substance is, by modern botanists, known to exist in several plants besides this, as in the stems of grasses and in the Horse-

tails; and the geologist finds the finest tissues
of bygone ages most perfectly preserved by its
means.   Long before these discoveries, however, the hardness of the Gromwell seeds had
been remarked, and significant names bestowed
on the plant in various countries.   The botanist
names it from *Lithos*, a stone, and *Spermum*,
seed; the English Gromwell is from *Graun*,
the Celtic for seed, and *mil*, stone.   It is the
Steinsame of the Germans, and the *Steenzaad*
of the Dutch.   Country people also call it
Grey Millet.

The flowers of the Gromwell are of a
yellowish white, blooming in June.   The plant,
though rare in Scotland, is common in England
by road-sides and on other waste places.   Its
stem is branched, and about a foot or a foot
and a half high.

There are three other British species.   One
is very frequent in the corn-field, and is very
similar to the plant figured here.   It is the
Corn Gromwell (*Lithospermum arvense*).   The
others are rare plants.

MUSK THISTLE.—*Carduus nutans.*

Class SYNGENESIA. *Order* ÆQUALIS. *Nat. Ord.* COMPOSITÆ. COMPOUND FLOWERS.

THE thistles are among the handsomest wild-flowers of the northern hemisphere, and some of them rise up before us by every way-side. Those who are not botanists can always detect these plants from others whose flowers some-what resemble them, by the prickly stems and leaves which always belong to the thistles; but some study of plants is requisite to distinguish the various species from each other. The Musk Thistle, however, may at once be known by its large drooping flower, and it has besides a musky odour, which becomes stronger when the dew of evening is on it. The colour of the blossom is a rich reddish purple, and it nods, from a stem two or three feet high, on many a dry or stony field, during the months of July and August.

Thistles are arranged by botanists into several genera; that of *Carduus* contains four species, one of which, the Welted Thistle (*Carduus acanthoides*), is among the most common of

the whole tribe; while another, the Milk Thistle
(*Carduus Marianus*), is as handsome as any
one of this beautiful family of plants, but it
is rare. It may easily be distinguished by the
milky white veins which run through its dark
green spiny leaf. The name of the genus is
said to be from the Celtic *Ard*, a point; and
our engraving will show that it was not ill
bestowed. This thistle, however, has not,
except on its cup, points so strong and sharp
as those of the true Scotch thistle (*Onopordum
Acanthium*), which certainly well merits the
motto which Scotsmen of old have affixed to
their national emblem, "Nemo me impune
lacessit" (No one touches me with impunity):
or, as Baxter interprets it into the plain
Scotch, "Ye maun't meddle wi' me."

LESSER DODDER.—*Cuscuta Epithymum.*

Class PENTANDRIA.   *Order* DIGYNIA.   *Nat. Ord.* CONVOLVULACEÆ.
BINDWEED TRIBE.

THE waxy looking flowers of the Dodder
are often seen by those who ramble on heaths
or open downs during the months of August
and September. This is one of our few native
parasitic plants, and it grows especially on
the Furze, but also on the Heath, Thyme,
Vetches, Trefoils, and other smaller plants,
entwining them with its long leafless red
threads, and hanging its clusters of small
pinkish-white blossoms among their foliage.
The seed of the Dodder takes root in the
soil, but puts forth a spiral shoot, which soon
winds about the neighbouring plant, and dis-
connecting itself from the earth, derives its
nutriment from the living juices. When
prevailing to any great extent, it proves very
injurious to the plants on which it is para-
sitical. It often forms a dense network over
the Furze, and Mr. Dovaston remarks, that he
once saw it in such profusion at Liphook in
Sussex, that it absolutely pulled down and
killed the Nettles. Baxter remarks of another

less frequent species, *Cuscuta Europæa*,— "About twenty years ago, I saw at Cassington, near Oxford, a large field of beans completely matted together with this parasite. It had taken possession of the whole crop, and having elevated itself several inches above the beans, produced a beautiful effect, especially when the sun shone upon it." Crops of Lucerne have been in France and Italy much injured by the larger species, which has thicker stems than the more frequent kind, and its flowers are of a yellowish rose colour. A third species, called the Flax Dodder (*Cuscuta Epilinum*), is sometimes most destructive to the Flax crops; and the Trefoil Dodder (*Cuscuta Trifolium*) is so troublesome in many places on Clover, that in some countries the farmer is actually prevented from attempting the culture of that plant. The Dodder boiled with ginger is a favourite village medicine, but the parasite is much disliked in country places. It is sometimes called Red Tangle, but it has, besides, many names expressive of the disgust entertained for it by country people.

COMMON GOLDEN ROD.—*Solidago Virgaurea.*

Class SYNGENESIA. *Order* SUPERFLUA. *Nat. Ord.* COMPOSITÆ. COMPOUND FLOWERS.

IF we would know something of the value of this flower to the insect race, we should watch it for some minutes on a sunny day of September or October. Our latest butter-flies and moths, our wild and hive bees, hover about it in numbers, or are settling in crowds on its golden blossoms. Flowers are few at this season, and as this will grow on the worst of soils, and needs no care, it would be wise of those who keep bees to plant it near the hives. It has been said, that an acre of arable land planted with it, would furnish at the close of the season a sufficiency for a hundred hives to complete their winter stock. Few plants vary more than the Golden Rod in mode of growth. If we find it in the dry wood, its flowers are usually small, and scattered on the stem, and often of a lighter colour. But gather a specimen from the mountainous pasture, or from the seaside cliff, where it is also plentiful, and we find it with a shorter, stouter stem, and with crowded clusters of flowers like those of our engraving.

This plant is called also Aaron's Rod and Woundwort, and is the only British species. The genus received its name from *solidare* (to unite), because of the supposed use of some of the Golden Rods in healing wounds; and as Gerarde says, large quantities of this very plant were, in his time, brought from beyond seas, and sold as a vulnerary. Our native flowers were then less known than now, and our countrymen had but recently discovered that the Golden Rod grew wild. Our good botanist recorded, in his " Herbal," his out-pourings of wrath against the disuse into which it was falling. He says, " It may truly be said of fantastical physitions, who when they have found an approved medicine and perfect remedy neere home, against any disease, yet not contented with that, they will seeke for a new farther off, and by that means many times hurte more than they helpe." He adds, that he has said this in order that these " new-fangled fellows " may be brought back again to esteem this admirable herb; but Gerarde's admonitions seem to have availed nothing.

COMMON RAGWORT.—*Senecio Jacobæa.*

Class SYNGENESIA. *Order* SUPERFLUA. *Nat. Ord.* COMPOSITÆ. COMPOUND FLOWERS.

THIS is an autumnal flower, and though it opens in July, yet in November it is still the companion of the yellow leaf, and of the tune of the bird which tells of coming winter.

" My childhood's earliest thoughts are link'd with thee,
The sight of thee calls back the robin's song,
   Who from the dark old tree
Beside the door, sang clearly all day long,
   And I, secure in childish piety,
Listen'd as if I heard an angel sing,
   With news from heaven, which he did bring
Fresh every day to my untainted ears,
When birds and flowers and I were happy peers."

The rich golden clusters of the Ragwort are very handsome, and growing on their tall stems, sometimes two feet in height, large quantities of the plant form a very attractive feature of the landscape. But it is very troublesome in the pasture, for its downy balls of seeds are scattered by winds and rains, and produce it in great abundance. This species grows both on dry and moist lands, but a very similar kind, the Marsh Ragwort (*Senecio aquaticus*), is peculiar to the margins of rivers and ditches, or the wet lands near them. Its leaves are less divided, and the lower ones are entire, while the flowers are usually larger than the Common Ragwort. It is in bloom at the same season.

Besides this, there is on chalky and gravelly soils a species less general than either of these, and easily distinguished from them. It is the Hoary Ragwort (*Senecio tenuifolius*), and its stems and the under surfaces of its leaves are quite white with cottony down. There are besides four rarer kinds of Ragwort.

The two species commonly known as the Groundsels, are familiar to every eye. The Common Groundsel (*Senecio vulgaris*) flowers all the year long. It is valuable as food for the caged and wild birds, and was in former days prized as a medicine both for man and animals. The Mountain Groundsel (*Senecio sylvaticus*) is a larger species, which is common on gravelly soils, flowering from July till September. Its leaves are more deeply divided than the former kind, and are sometimes hoary.

The silky down of the seeds, so like the silver hair of age, suggested the name of *Senecio*, from *Senex*, an old man.

DEPTFORD PINK.—*Dianthus Armeria.*

Class DECANDRIA. *Order* TRIGYNIA. *Nat. Ord.* CARYOPHYLLEÆ.
PINK TRIBE.

OUR native Pinks are comparatively rare
flowers. This species is the most frequent, but
though found on many pastures, both of
England and Scotland, it is by no means
general. Its rose-coloured petals are dotted
with small white spots, and the flowers grow
in clusters on very long stems. These are
usually one or two feet high, and the specimen
from which the drawing was taken, was nearly
a yard in height. The flower is quite scentless,
and blooms in July and August. One other
of our wild Pinks only bears its flowers in
clusters. This is the Proliferous Pink (*Dianthus prolifera*), which has in July small and
deeply coloured blossoms, of which only one
in the cluster opens at a time. The flowers
are rose-coloured, and the species is distinguished by the brown dry scales in which they
are enclosed. It is a rare plant of gravelly
pastures.

We have three wild kinds of Pink with
flowers growing singly, or with only two or

three on the stem.   One of these, which is
large and deliciously fragrant, is interesting
as being the origin of the garden Carnation.
Though less beautiful than that flower, it
resembles it in scent.   It is the Clove Pink,
or Clove Gillyflower (*Dianthus caryophyllus*),
and is also commonly termed Castle Pink, as
it grows on old castles and walls.   Deal, San-
down, and Rochester castle, as well as an old
wall at Norwich, are the well-known places on
which this beautiful flower springs up among
the remains of the past.   Like some others
of the ruined wall, as the Snapdragon, it is
but a doubtful native, but we may see it either
in its red or white varieties, as a common
ornament of the cottage garden.

The Mountain Pink (*Dianthus cæsius*) is a
rare but handsome flower of the limestone
rock; and the Maiden Pink (*Dianthus del-
toides*), which is a rather better known kind,
grows on a stem about a foot high, and has a
small rose-coloured flower dotted with white,
with a white eye enclosed in a deep purple
ray.   It grows on gravelly pastures.

COMMON BUR-DOCK.—*Arctium Lappa.*

Class SYNGENESIA. *Order* ÆQUALIS. *Nat. Ord.* COMPOSITÆ.
COMPOUND FLOWERS.

THIS large and branched plant is common everywhere in the country, by waste places and in woods and hedges, and is made conspicuous by the large wavy leaves around its root. The lilac, or deep purple blossom, appears in July and August, and crowns a globose flower-cup, which every schoolboy recognises as the burr, so frequently used in sport, and too often wantonly thrown at the bats to bring them to the ground. This burr is a ball of hooked spines, and the whole structure is evidently adapted to the dissemination of the seeds, which, adhering to the wool of animals, or the clothing of man, are thus spread far and wide. The plant is familiarly called Hurr-burr, Clot-burr, and Great-burr, and in Britain, where it is kept down by cultivation, it is seldom an annoyance; but in the vast uncleared lands of Australia, similar plants are very troublesome; and the Burr of the colonist, which like this has a prickly ball of seeds, attains a great size and causes much inconvenience by adhering to the stockings or other parts of the clothing.

Several intelligent physicians have been of opinion, that a decoction of the roots of our Bur-dock are not inferior in medicinal powers to Sarsaparilla.   This decoction, as well as the powdered seeds, are frequently used with success as a remedy for rheumatism, and even the leaf laid on the part affected is considered efficacious.   The author has many times been assured by villagers, that their rheumatic pains were cured by this outward application; but as the popular notion of the relation between cause and effect is not always the philosophical one, this mode of applying the plant must not be relied on too confidently.   When burnt before flowering, the Bur-dock yields a good quantity of alkaline salt, equal to the best potash.   The young shoots may be boiled like asparagus, or eaten uncooked with oil and vinegar.

COMMON HENBANE.—*Hyoscyamus niger.*

Class PENTANDRIA.   *Order* MONOGYNIA.   *Nat. Ord.* SOLANÆ.
NIGHTSHADE TRIBE.

THOSE who have been accustomed to observe flowers would, even at first glance, suspect that the Henbane was poisonous. Its dingy cream-coloured blossoms, veined with a lurid purple, would confirm the impression made by the disagreeable and sickly odour of the whole plant.   Dull yellow, dim purple, or green flowers, often characterise noxious plants, though these distinctions are not invariable. The Henbane is powerfully narcotic, especially at that time when the flowers have just fallen, and the seeds are ripening. Lightfoot mentions that a few of these seeds have been known, when eaten, to deprive a man of his reason and the use of his limbs ; though both Professor Martyn and Sir J. E. Smith ventured to eat them, and did so without injury.   Few animals will touch it.   The sheep will sometimes feed sparingly upon it; and the goat, which seems to have a wonderful power of feeding safely on the most unwholesome plants, will also eat this.   Swine, too, are said to relish it; and

as its seed has some slight resemblance to a
bean, the scientific name signifies Hog's-bean,
and the plant in many country-places is so
called. The foliage and stems of the Henbane
are very downy and viscid, and its flowers grow
in clusters among the leaves during June and
July. It is a common plant of waste places,
and not unfrequent in churchyards. Though
its juices are poisonous if improperly administered, yet the Henbane yields a valuable
narcotic medicine, producing sleep without
those restless and distressing symptoms which
succeed the use of many medicines of this
nature. It is sometimes smoked like tobacco
by country people as a remedy for toothache,
but convulsions have occasionally followed its
use in this way.

FRAGRANT LADY'S TRESSES.—*Neottia spiralis.*

Class GYNANDRIA. *Order* MONANDRIA. *Nat. Ord.* ORCHIDEÆ.
ORCHIS TRIBE.

THE English name of this plant recals the times when the monks were the chief observers of flowers, and named so many in honour of the Virgin Mary and the different saints. All the flowers having the prefix of Lady, or the word Mary, were doubtless called after the Virgin, and were originally Our Lady's flowers. Hence we have Lady's Bedstraw, Lady's Mantle, the Rosemary, and the Marygold. The Costmary is derived from *Costum Mariæ*, or Mary's Balsam; while the Sweet Basil, Sweet Cicely, and many others took their names from those who reverenced the saints of the Romish Calendar. Our Lady's Tresses shares its name, apparently, with another plant; for dried branches of sea-weed are in Cornwall affixed to wooden stands, and placed on the chimney-piece. They are called Lady's Trees, and the poor people believe that they protect the house from the danger of fire. A learned writer, commenting on this superstition, remarks, "We believe that Lady's Trees is a

corruption of Lady's Tresses, a name still borne
by a flower not uncommon on our chalk downs,
and which seems to be an old one; for the
twisted spike of that flower resembles one of
the hair dressings in use at the close of the
fourteenth century."

This twisted spike of blossoms is a singular
characteristic of the pretty little orchis, which
in September and October blooms on the
chalky hills of England; making little show
there, however, by its small greenish-white
flowers. It is from four to six inches high, its
blossoms arranged in a single row, and turning,
in some specimens from left to right, in others
from right to left, round the upper part of the
stalk. The leaves grow in little tufts just above
the crown of the root, and do not appear till
the flower has begun to expand. The Rev.
C. A. Johns observes in his " Flowers of the
Field," that they are remarkably tenacious of
life, continuing to unfold even while subjected
to the pressure made on the blotting-paper
when drying for the herbarium.

YELLOW WEASEL-SNOUT.—*Galeobdolon luteum.*

Class DIDYNAMIA. *Order* GYMNOSPERMIA. *Nat. Ord.* LABIATÆ. LABIATE TRIBE.

THIS Yellow Archangel, as it is sometimes called, is very similar in its mode of growth to the common White Dead Nettle, which in summer time may be found under almost every hedge. We wonder not that botanists, from Gerarde downwards, often classed it with that flower, though it is now placed in a distinct genus. It is generally taller and more slender than the White Nettle, often growing to the height of a foot and a half. The whorls of flowers are very handsome, being of bright yellow, mottled with reddish orange. Its scientific name, made from two Greek words, signifies that it has the scent of a weasel, and certainly its odour is very disagreeable, especially if the plant be bruised. The Dutch call it by a name signifying Dog's Nettle, and the French term it *L' Ortie des Bois.* It blossoms in May and June, and though not a common flower in many counties, the woods of some parts of Kent are full of its bright blossoms. The plant can make small claim to medicinal

virtues, though it is slightly astringent; yet
few who saw it growing in large numbers
would fail to recognise its beauty, and to feel
that the beauty of flowers has its uses too.
Thus Mary Howitt has said,—

> " Springing in valleys green and low,
>   And on the mountains high,
>   And in the silent wilderness
>   Where no man passeth by:

> " Our outward life requires them not;
>   Then wherefore had they birth?
>   To minister delight to man,
>   To beautify the earth,

> " To comfort man, to whisper hope
>   Whene'er his faith is dim;
>   For Whoso careth for the flowers,
>   Will care much more for him."

The Yellow Weasel-Snout is the only British
species, and it is a native also of France,
Germany, Switzerland, Austria, Carniola,
Italy, and Holland. It is common in some
parts of Sweden, and the most northern limits
known for the flower are Wasa in Finland,
and Drontheim in Norway. There is a very
pretty variety of this plant, with variegated
leaves.

HAWKWEED PICRIS.—*Picris hieracioides.*

Class SYNGENESIA. *Order* ÆQUALIS. *Nat. Ord.* COMPOSITÆ.
COMPOUND FLOWERS.

As the first gleam of spring sunshine is reflected by the stars among the grass, the silver daisy and the golden dandelion, so the fainter ray of the winter sunbeam falls on a starry flower—a star of gold. This Picris is in bloom from July to October, but when the season is mild, it may be seen among the grass or the withered leaves, both at the close of the old year and the dawn of the new, with its yellow tint as bright as ever. We might say to this flower as Lowell did to another:—

" How like a prodigal doth Nature seem,
   When thou, for all thy gold, so common art!
Thou teachest me to deem
   More sacredly of every human heart,
Since each reflects in joy its scanty gleam
Of Heaven, and could some wondrous teaching show,
Did we but pay the love we owe,
And with a child's undoubting wisdom look
On all these living pages of God's book."

This flower is very abundant, growing to the height of two or sometimes three feet, on the borders of fields, by roadsides, and on hedge-banks. It is rather slender, and its stem is rough with hooked bristles. The blossoms are very numerous, and are chiefly on the branches at the upper part of the stem. In autumn, the hairy leaves are generally edged with a dark line of red.

This Hawkweed Picris is the only British species of the genus, but some handsome kinds, introduced hither chiefly from the countries near the Mediterranean, are very ornamental border flowers of the garden. The genus has very slight medicinal properties, but as some of the plants possess a small amount of the bitter principle, it was named from the Greek *Picros*, bitter. Our common flower is almost as frequent a one on the roadsides of nearly every country of the continent, as on ours. The Germans term it *Das Bitterkraut*, and the Dutch have the synonymous name of *Bitter-kruid*. In the olden times, it was familiarly called in England, Yellow Succory, and Yellow Ox-tongue.

RED HEMP NETTLE.—*Galeopsis Ladanum.*

Class DIDYNAMIA. *Order* GYMNOSPERMIA. *Nat. Ord.* LABIATÆ. LABIATE TRIBE.

THIS is by no means an uncommon plant during the months of August and September, in gravelly and chalky corn-fields. The blossoms of purplish rose-colour mottled with crimson, are, except the Mayweed and a few yellow starry flowers, almost the only bright things left among the stubble of the field, whence the brown corn has been gathered in, and on which the brilliant cornflowers have fallen with it, or died away by its side.

This Hemp Nettle has a stem from ten to twelve inches high, but an equally, or indeed more frequent plant, is that termed the Common Hemp Nettle (*Galeopsis Tetrahit*), which is somewhat similar to this species, but has broader leaves, and is altogether much larger, its stem usually attaining one or two feet in height. Its flowers are variegated with light purple and yellow, and are often nearly or quite white. They grow in whorls around the stem, and are remarkable for the long sharp teeth of their calyxes. This species may at once be known from the red kind by the circumstance that the stem is swollen beneath every pair of leaves. It blossoms in corn-fields and on cultivated grounds from July to September.

Two other British species belong to this genus, but they are not common plants. The large-flowered Hemp Nettle (*Galeopsis versicolor*) blooms in July and August in the corn-fields of Norfolk, and some few other places. It resembles in general appearance the common Hemp Nettle, but has much more beautiful flowers, their bright yellow tint being varied by a broad purple spot on the lower lip. It has rank foliage, and grows two or three feet high. Though so rare in England it is very common in Scotland, especially in the Highlands.

Rare also is the Downy Hemp Nettle (*Galeopsis villosa*), which is more like the Red Hemp Nettle, though its flowers are of a pale yellow. It grows in sandy corn-fields in Yorkshire, Lancashire, and Nottinghamshire, and also at Bangor in Wales.

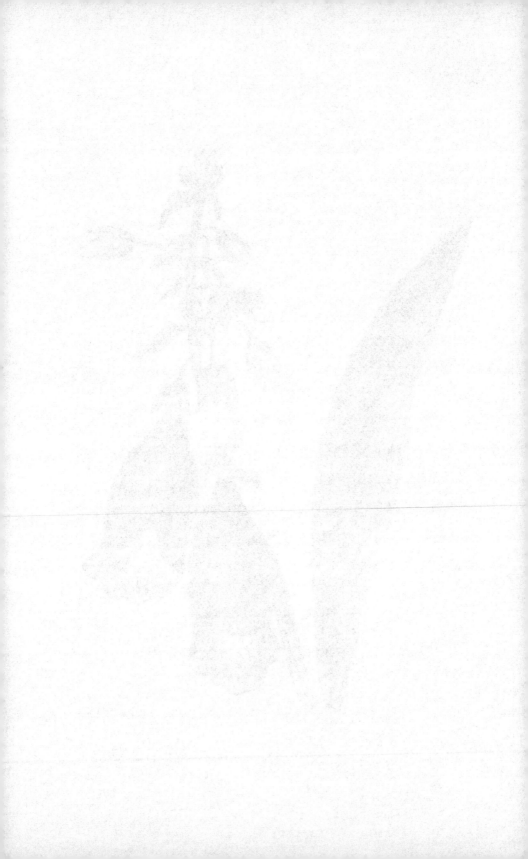

PURPLE FOXGLOVE.—*Digitalis purpurea.*

Class DIDYNAMIA. *Order* ANGIOSPERMIA. *Nat. Ord.* SCROPHU-
LARINEÆ.

FIGWORT TRIBE.

THE Foxglove, with its active properties and its stately form, has long been the

"Emblem of Cruelty and Pride,"

and is one of the handsomest of our native plants. Growing commonly to the height of three or four feet, and in some fine specimens being even six feet high, it is so conspicuous as often to be taken by the painter for the ornament of the foreground of a rural landscape. The blossoms, of which our small page can represent so few, are very numerous, and sometimes occupy nearly two feet of the length of the stem. The Foxglove is common on rocky and hilly districts, and from Carrington's poem appears to be very abundant in Devonshire.

"Upon the sunny bank
The Foxglove rears its pyramid of bells,
Gloriously freckled, purpled and white, the flower
That cheers Devonia's fields."

But in many Eastern counties of England, as in Norfolk and Suffolk, this plant is almost unknown. It blossoms in June and July, and its leaves are large and veiny. The flowers are sometimes white, and they are beautifully speckled within.

The Digitalis has long been valued for properties which, though poisonous when administered too freely, yet when skilfully used, give most important relief to the sufferer. The foliage is the part employed in medicine, and unless it be dried by a quick process shortly after being gathered, it loses its power. On this account the country practitioner generally prepares it himself, and we can remember many long summer walks taken during childhood, over Kentish hills and dales, to procure it for a friend.

The scientific name is from *digitali*, the finger of a glove. The Germans term it *Fingerhut*, and the Dutch *Vingerhoed*. In France it is called *Doigts de la Vierge* and *Gants de Notre Dame*. Our Foxglove is thought to be a corruption of " Folk's glove," that is, Fairies' glove.

PERENNIAL, OR DOG'S MERCURY

PERENNIAL, OR DOG'S MERCURY.
Mercurialis perennis.

Class DIŒCIA. *Order* ENNEANDRIA. *Nat. Ord.* EUPHORBIACEÆ.
SPURGE TRIBE.

EVERY one at all used to gardening knows
this weed, for, no sooner is a plot neglected
than it is overrun by it. Gardeners near the
metropolis, as well as in other large cities, are
as much annoyed with it as those in the
country; hence it is known in some places, as
at Dovor, by the name of Town Weed. It is
also abundant in fields and woods, and by
almost every road side.

This plant was in former times commonly
called Dog's Cole, both that and its present
name referring to the idea that it was eaten by
dogs. It is called almost throughout Europe
by some word synonymous with Mercury, from
the notion that supposed medicinal virtues in
the plant were revealed to the world by him;
the French terming it commonly *La Mercuriale,*
and the Italians *Mercorella.* As might be
inferred from its appearance, this herb is of a
poisonous nature. Ray records an instance of

a family who suffered much from its deleterious effects, after having eaten it with fried bacon.

The barren flowers, which are green, grow on long slender spikes. The fertile ones are much less conspicuous, and nestle among the upper leaves. They must be sought for on separate plants. When drying for the herbarium, it loses all its green colour, and becomes of a bluish or black hue. A good blue dye may be procured by steeping the plant in water, but no means hitherto known have availed to render it permanent. It grows to the height of a foot or a foot and a half, and is in flower from April to the end of summer.

The only other British species is the Annual Mercury (*Mercurialis annua*), which grows in waste places about towns and villages, but is not frequent. It is about a foot high, and bears longer spikes of green flowers.

COMMON TANSY.—*Tanacetum vulgare.*

Class SYNGENESIA. *Order* SUPERFLUA. *Nat. Ord.* COMPOSITÆ. COMPOUND FLOWERS.

EVERY cottage garden contains the fragrant Tansy, for there, with

"Marjoram-knots, Sweet-briar, and Ribbon-grass,"

its dark green leaves wave on the little bed enclosed by daisies. But the Tansy is a wild flower too, and is common on hedge-banks and waste places, easily distinguished by its clusters of yellow button-like blossoms, and its powerful and aromatic odour.

"Fragrant the Tansy breathing from the meadows,
As the west wind bows down the long green grass;
Now dark, now golden, as the fleeting shadows
Of the light clouds pass, as they wont to pass
A long while ago!"

The Tansy often flourishes well by the sides of rivers, as on the banks of the Avon, and thrives too on sea-banks, as on those of Sandgate in Kent. It is sometimes collected by country people to make Tansy wine, which is thought to have some valuable remedial effects, and in Scotland a decoction of the plant is used as a medicine for gout. Meat rubbed with Tansy is said to be preserved from the injury of flies, but its strong flavour would certainly not improve that of the meat. The bitter herbs commanded in ancient days to be eaten at the Paschal season are typified in the Romish Church by this plant. Cakes made of eggs and the young leaves of this flower, and called Tansies, are now eaten during Lent; and these and the Tansy puddings, which are also made of the herb, are sufficiently nauseous and bitter to be eaten by way of penance. Many country people, however, eat the puddings with much relish.

The Common Tansy is the only British species. It flowers during August and September, and grows to the height of two or three feet. It is a perennial plant. The name of *Tanacetum* is altered from the Greek *Athánaton*, everlasting: and our Tansy is a corruption of *Athanasia*, "as though," says Gerarde, "it were immortal, because the flowers do not easily wither away."

PRIVET.—*Ligustrum vulgare.*

Class DIANDRIA. *Order* MONOGYNIA. *Nat. Ord.* OLEACEÆ.
OLIVE TRIBE.

THE small white flowers of this plant, arranged in their dense pyramidal clusters, are very common in our hedges during May and June. In the town garden, too, we often see them among the green boughs which form a hedge for its enclosure, or cover the arbour. The Privet grows more rapidly than the Whitethorn, and no native shrub bears so well as this the smoke of the city, while the dripping of trees does not injure it. Its roots are short and slender, thus occupying a smaller portion of soil than most other shrubs, and when cultivated it is an evergreen. In its wild state many of the leaves fall off in winter, but some few remain on the boughs till the following spring, and the dark purple berries stand in handsome clusters, and serve as winter food to birds when berries are scarce. The bullfinch especially feeds upon them in the cold season. Though this plant is remarkable as being little liable to injuries from the insect race, yet the Privet Hawkmoth, while in the caterpillar

state, feeds chiefly upon it; and the Blister
Beetles, or Cantharides, revel among its
verdure.

The Privet grows throughout Europe. Its
old name was Primprint, which Professor
Martyn conjectures it owes to its bearing the
trimming shears of the gardener so well.
Perhaps the somewhat formal and neat appearance both of its flowers and leaves might
have suggested its name. The berries of the
shrub dye silk or woollen material a beautiful
and durable green. The leaves are slightly
bitter in flavour, and the wood is hard and of
some use to the turner. It is easily propagated
by layers and suckers, but the best plants are
those raised from seed. The Spaniards and
Portuguese call it *Al Hena*, which is the name
of the Oriental dye by means of which the nails
are stained pink. Perhaps our plant may have
received its name in consequence of a rose-
coloured pigment which may be procured from
the leaves.

This Privet is the only British species of
the genus.

UPRIGHT SPIKED SEA-LAVENDER.
Statice binervosa.

Class PENTANDRIA. *Order* PENTAGYNIA. *Nat. Ord.* PLUMBAGINEÆ.
THRIFT TRIBE.

FEW wild flowers are better adapted than this to form a bouquet which shall last throughout the winter. In towns and villages which are near our rocky coasts or salt marshes, we often see large nosegays of these Lavenders mingling with dried grasses, to form an ornament for the winter chimney-piece, or the summer stove. The species here figured, though most plentiful on Dovor cliffs and some other rocky places near the sea, is not so generally distributed as the Larger Sea-Lavender of the muddy shore or salt marsh. Our engraving represents but a small plant, many specimens being a foot and a half in height. The leaf sufficiently resembles the spatula of the druggist to have rendered its former name of Spathulate Sea-Lavender quite appropriate, and at once distinguishes this from the other species. The delicate blue-lilac blossoms appear in August, and the stems and foliage are of bluish green. This plant is much sought after by sea-side visitors, who carry it to their homes as a remembrance of pleasant hours spent among sea-cliffs and sounding waves. Happily, much of it grows on steep acclivities beyond the reach of the wanderer, or the plant might soon be altogether extirpated by the inconsiderate manner in which such quantities are carried off. Though the blossom resembles in colour that of the garden plant of the same name, yet it is but—

" The Sea Lavender, which lacks perfume."

The more frequent species, the Spreading Spiked Sea-Lavender (*Statice Limonium*), is very similar to this, but is larger, and the blossoms grow more at the top of the stem. The leaves, too, are very different in size, varying from that of about four inches to a span, and placed on stalks. It blossoms in marshes in July and August, and like this flower retains its form, and something of the blue tint of its flowers, long after it has been gathered.

The Matted Sea-Lavender (*Statice reticulata*) occurs only on the salt marshes in Norfolk, and its flower-stems are divided almost from the base of the plant. It is much smaller than either of the preceding species.

PLOUGHMAN'S SPIKENARD.

PLOUGHMAN'S SPIKENARD.—*Inula Conyza.*

Class SYNGENESIA. *Order* SUPERFLUA. *Nat. Ord.* COMPOSITÆ. COMPOUND FLOWERS.

THIS plant has a very dull appearance, with its dark green leaves and dingy yellow flowers, which, from the shortness of their rays, look as if not yet fully expanded.

It is common on chalky and limestone soils in England, either in hedges or on exposed places, as on the cliffs at Dovor, where it is very abundant. It is a rare flower, however, in Scotland. The stem of this Spikenard is often two or three feet high, and many leaves grow among its panicle of blossoms, the lower ones being stalked. While growing, it does not diffuse any odour; but on bruising the plant, the scent is very strong, and is by some persons thought agreeable. Clare, in his beautiful lines to Cowper Green, alludes to it :—

> " And thou hast fragrant herbs and seed
> Which only garden's culture need ;
> Thy Horehound tufts, I love them well,
> And Ploughman's Spikenard's spicy smell ;
> And Thyme, strong-scented, 'neath one's feet,
> And Marjoram buds, so doubly sweet,
> And Pennyroyal's creeping twine,
> These, each succeeding each, are thine."

This plant, as well as several others, is known in some country-places by the name of Flea-bane, and in France by the synonym of *Herbe aux puces*, and it is supposed to be noxious to some insects. It blossoms in July, and continues in flower till October, after which, the calyxes are full of little tufts of seeds surmounted with down.

Two other species of Inula are British plants. The large yellow flower called Elecampane (*Inula Helenium*) is a rare plant on the moist pasture, and is possessed of some valuable medicinal properties. The other, the Golden Samphire (*Inula crithmoides*), is a very unfrequent plant, but it grows on some salt marshes and sea cliffs on the south-west shores of England, as well as on some parts of the Welsh coast. It is about a foot high, and blooms in August, each branch bearing a single yellow flower. The name of the genus is supposed to be a corruption of *Helenula*, Little Helen, the celebrated Helen having, as tradition tells, improved her beauty by a cosmetic made from the species called Elecampane.

WOOD BETONY.—*Betonica officinalis.*

Class DIDYNAMIA. *Order* GYMNOSPERMIA. *Nat. Ord.* LABIATÆ. LABIATE TRIBE.

THE Betony has some general resemblance to the Dead Nettles and Woundworts, but it has one peculiarity by which it may be known from all other of the red Labiate flowers. Its spike is formed of two portions. The upper portion comprises a number of whorls, then we see a small piece of the stem without any flowers, while the remainder of the spike, formed of a small number of whorls, is quite distinct from the upper part, and has two leaves at its base. The Betony is common in woods and thickets, and grows to the height of one or two feet : it has but few leaves, and the lower ones grow on long footstalks. The flowers appear in August, and are of a very rich purplish red. It is the only British species of the genus.

The fresh flowers of the Betony are said to have an intoxicating effect, and the dried leaves to cause sneezing. The plant is still gathered and dried for a medicine in country places, though we no longer prize it as it was prized in earlier days. The Romans had a proverb, " Sell your coat and buy Betony;" and another old proverb was, " May you have more virtues than Betony." Antoninus Musa, physician to the Emperor Augustus, wrote in high terms of its excellences, and stated that it would cure forty-seven disorders to which humanity was subject. Franzius, in his "History of Brutes," not content with the repute which Betony had obtained as a cure for human ills, cites it as of medicinal value to the lower animals. He says of the stag, " When he is wounded with a dart, the only cure he hath is to eate some of the herbe called Betony, which helpeth both to draw out the dart, and to heale the wound." Sir William Hooker thinks that the modern name is a corruption of Bentonic; " ben " meaning head, and " ton " good or tonic.

The root of this plant is bitter. The whole herb may be used to dye wool of a fine yellow colour. It is also smoked as tobacco in country places. The Germans call it *Die Betonika*, the Italians *Betonico*, and the French *La Bétoine*.

COMMON HOP.

COMMON HOP.—*Humulus Lupulus.*

Class DIŒCIA. *Order* OCTANDRIA. *Nat. Ord.* URTICEÆ.
NETTLE TRIBE.

THIS graceful twining plant hangs down its small loose clusters of fragrant cones on many of our hedges, in July and August. It is rather naturalized than indigenous, but has, for some centuries past, grown wild in some counties of England. The stems are long and trailing, and thickly set with small prickles. It is largely cultivated in many parts of England, that its catkins may be used to give the bitter flavour to beer; and our country hardly presents a more picturesque rustic scene than that of the hop-garden, during September or October, when men, women, and children are busily engaged in unwreathing the hop-poles. The culture of this plant was introduced into England from Flanders, in the time of Henry VIII., but our forefathers thought the hop unwholesome, and many strongly opposed its use. Before this period, various wild plants had been mingled in beer, as the Wood Sage, and especially the Ground Ivy, or Ale-hoof, as it was called. Ale was commonly drunk at a very early period, in this country:

"King Hardicanute, 'midst Danes and Saxons stout,
  Caroused in nut-brown ale, and dined on grout."

And one of the oldest Welsh songs says :—

"But we from the horn, the blue silver-rimm'd horn,
Drink the ale and the mead in our fields that were born."

The generic name of the Hop is derived from *humus*, rich earth, which is needful to its culture; and the English word seems to have come from the Saxon *hoppan*, to climb. It is a powerful narcotic, and one of the very best of vegetable tonics. The soothing power of its fragrance has induced physicians to recommend a hop pillow in cases of sleeplessness, a remedy which George III. was accustomed to use in illness. The young tops of either the wild or cultivated plant are tender, and when boiled, may be eaten like asparagus. Some of us have, perhaps, in childhood, thought them sweeter than any other vegetable. They were formerly brought to market in small bundles. The stems and leaves of the hop dye wool yellow, and a decoction of the root is said to be as beneficial a medicine as the sarsaparilla. A strong cloth is manufactured in Sweden from the fibres of the stalk.

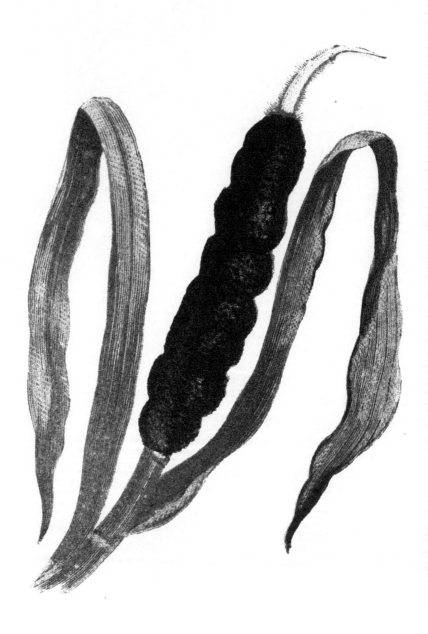

GREAT REEDMACE

GREAT REED-MACE, OR CAT'S-TAIL.
Typha latifolia.

Class MONŒCIA. *Order* TRIANDRIA. *Nat. Ord.* TYPHACEÆ.
REED-MACE TRIBE.

THIS plant, which is often incorrectly called Bulrush, is frequently represented by the Italian painters as placed for a sceptre in the hand of the Saviour, when, in mockery, the soldiers called him King of the Jews. It is our largest herbaceous aquatic, and our illustration can but represent some of the smaller leaves, many of these being three or four feet long. The stem is often six feet in height, and is surmounted by the olive-brown spike, varying from the size which we have represented, to one of a foot in length. The lower portion only of this contains fertile flowers, the upper ones being barren. This Reed-Mace is a handsome ornament to our river sides, and is equally conspicuous on those, not only of European countries, but of New Zealand, Australia, and of most parts of the world. Cottages in England, and the huts in uncivilized lands, often owe some of their covering to its long leaves, which form a good thatch;

and Kalm, who found it very frequent in North America, mentions a variety of uses to which it is there applied. In our country, mats and baskets are made of its leaves. Haller says that the roots are eaten in salads. The pollen of the flower is very abundant, and produces a flash of fire if a light is applied to it.

An unripe spike of this plant will, if gathered and put in a warm place, gradually ripen its seeds, which are covered with a soft down. The author once left one in a small room, and taking it up some days after, and inadvertently shaking it when the window was open, hundreds of these seeds floated about the room, settling themselves in the curtain and carpets, and some time elapsed before they could all be driven from their holds. This woolly substance is said by Kalm to have been formerly used in Sweden instead of feathers, in beds and cushions, but its tendency to adhere and form knots caused its disuse. The plant flowers in July and August.

The other native species, the Lesser Reed-Mace (*Typha angustifolia*), is less frequent.

ROUND-LEAVED SUNDEW.—*Drosera rotundifolia.*

Class PENTANDRIA.   *Order* HEXAGYNIA.   *Nat. Ord.* DROSERACEÆ.
SUNDEW TRIBE.

A VERY curious and pretty little plant is this Sundew, which is found on the bogs of many parts of our country during July and August. Its leaves are thickly covered with red hairs, each of which has a small drop of clammy fluid at its tip, so that even at noonday, and in brightest sunshine, the plant seems covered with the dew of evening. Insects are attracted by the sweet flavour of this fluid, and being entangled among the clammy hairs, die on the leaf.

A small stem, from two to six inches in height, bears on its top the few little white blossoms, but many persons who know the plant well have never seen the flowers fully open. Their time of expansion is from nine in the morning till noon; but we can never depend on their unfolding, for this seems in no way determined by the absence or presence of the sun, or by any obvious cause. The only way to make sure of seeing them is by gathering a plant in bud, and placing it in water. The author once examined, almost daily, during July, a number of these plants on their native bog, and never but once was fortunate enough to see them open. The Rev. W. T. Bree says, that he had known the plant from boyhood, and never saw it fairly unclosed till the year 1833. " Botanising the 27th of July last," says this gentleman, " on the shores of Coleshill Pool, where the Drosera grows in great profusion, I was surprised and delighted at beholding the flowers on every plant fully and beautifully expanded. The day was warm and bright, but the sunshine at intervals interrupted by passing clouds: the hour of the day must have been, I think, from a little before twelve till one." In the preceding summer this naturalist had visited the same spot, at about the same season, and could not find a single Sundew expanded.

A smaller species, the Long-leaved Sundew (*Drosera longifolia*), often grows with the former kind on bogs: and the rare species, called Great Sundew (*Drosera Anglica*), is occasionally found on similar places. All are acrid, and were formerly prized as medicines. A good cosmetic is made of their juices mingled with milk.

GREAT HAIRY WILLOW-HERB.—*Epilobium hirsutum.*

Class OCTANDRIA. *Order* MONOGYNIA. *Nat. Ord.* ONAGRARIÆ. WILLOW-HERB TRIBE.

IN wandering during July and August among lands through which the stream is winding, we shall be almost sure to see this flower, for it is a common plant by the side of our fresh waters, and is too large to be overlooked. It would seem at a distance almost like a small shrub, for the stem is often a yard high, and has many branches well clad with soft downy leaves, and with rich purple blossoms. The flowers have that mixture of red and dark blue in their tint, which it is difficult fully to describe, either by pen or pencil.

This flower is in many country places known by the name of Codlins-and-Cream, and Cherry-pie, because of its odour of boiled fruit. It is eaten by cattle, and an infusion of this, as well as of some other species, has an intoxicating effect. The down of some of these Willow-herbs, mixed with cotton, has been manufactured into stockings, and, mingled with the fur of the beaver, has served for hats

and other articles of clothing. There are
nine British species of this plant, but with
the exception of the Rose Bay Willow-herb
(*Epilobium angustifolium*), none of them are
so showy as the species we have represented.
The Rose Bay is frequent in Scotland, but is
not very common as a wild flower in England,
though frequently planted in gardens, under
the name of French Willow. The stems are
from four to six feet high, the flowers of a
reddish purple, and the pods so full of downy
seeds, that if once admitted into a garden, it
can with difficulty be extirpated. It grows
in some moist places; and in the wild woods
near Wrington, in Somersetshire, whole acres
are gay in August with its showy blossoms.
The young shoots are boiled and eaten, and
the pith, when boiled and properly prepared,
makes both good beer and vinegar; but the
whole plant has, when infused, a very intoxicating quality. The name of the genus, taken
from the Greek *epi*, upon, and *lobos*, a pod, is
very significant, as all the flowers are seated
on long pods.

GREATER KNAPWEED.

GREATER KNAPWEED.—*Centaurea Scabiosa.*

Class SYNGENESIA. *Order* FRUSTRANEA. *Nat. Ord.* COMPOSITÆ. COMPOUND FLOWERS.

THIS handsome bright purple flower is placed on a branched stem, two or three feet high, and is very common on barren pastures, roadsides, or corn-fields. Its season of blossom is July and August, but an occasional flower enlivens the winter, even till Christmas. The hard oblong calyx reminds us of its familiar name of Iron-weed; and Knap-weed was probably Knob-weed in former times. Clare enumerates these plants among the flowers of the corn-field.

> " Each morning now the weeders meet
> To out the thistle from the wheat,
> And ruin in the sunny hours
> Full many a wild-weed with its flowers;
> Corn-poppies, that in crimson dwell,
> Called Head-aches for their sickly smell ;
> And Charlocks, yellow as the sun,
> That o'er the corn-fields quickly run ;
> And Iron-weed, content to share
> The meanest spot that spring can spare :
> E'en roads where danger hourly comes
> Are not without its purple blooms ;
> Whose leaves, with threatening prickles round,
> Thick set, that have no strength to wound,
> Sink into childhood's eager hold,
> Like hair."

The Black Knapweed (*Centaurea nigra*) is an equally common flower, and is easily known by its smaller blossoms, which are also of much less vivid purple colour. Its leaves are rough, with numerous little bristles; those at the root are lyre-shaped, but those on the dark-brown stem are long and slender. The scales of the flower-cup are almost black, but thickly set round with brown teeth. This flower blossoms in July and August, and is popularly known by the name of Hard-head. It is a harsh, stubborn weed in meadows and pastures, very difficult of extirpation, and seldom eaten by cattle, either in its green or dried state. The expressed juice of the florets, both of this and the Greater Knapweed, are said to make a good ink.

The Brown-rayed Knapweed (*Centaurea Jacea*) grows on waste places in Sussex, and is abundant in Angusshire. It is a very handsome plant, with purple spreading florets. There are three other native species of *Centaurea*, the Corn Blue-bottle, and the two kinds of Star Thistle.

COMMON HEMLOCK.—*Conium maculatum.*

Class PENTANDRIA. *Order* DIGYNIA. *Nat. Ord.* UMBELLIFERÆ.
UMBELLIFEROUS TRIBE.

THE Hemlock may be known from plants which are very similar by its smooth stem, spotted with purplish brown, and its fetid smell. The little leaflets at the base of its small clusters, and which are called bracts, only go half way round the stalk, and the foliage is large and beautifully formed. The Hemlock has most powerful properties for good or ill, and when bruised the disagreeable odour is greatly increased. It may be found very frequently in July bearing umbels of white flowers, on banks, at the base of walls, and in other waste places. The poisonous nature of the plant was known both to the poets and philosophers of olden times, every part of it possessing, especially while fresh, a volatile oily alkali, termed Conia, which is so poisonous that a few drops will kill a small animal in a few minutes. Happily for us, men of science know more of the worth and responsibility of human life than to agree with Pliny as to the object for which such plants as the Hemlock were created. "Wherefore," says this old writer, "hath our Mother Earth brought out poisons in so great a quantity, but that men in distress might make away with themselves?" This plant, however, like the Foxglove, is adapted to the necessities of man, and is by means of human skill made to lengthen life, and not to shorten it, for it affords a valuable relief in nervous and other disorders. The injudicious use of Hemlock in villages has, however, often been attended with serious results. Gerarde says that the Common Marjoram, given in wine, is a remedy not only against "the bitings and stingings of venomous beasts," but that it also "cureth them that have drunke opium or the juice of blacke poppy or hemlocks, especially if it be given with wine and raisons of the sunne."

The Common Hemlock is the only British species. It is usually about three or four feet high, but is sometimes twice that height. The hollow stems are by country people called "kecksies."

IVY-LEAVED BELLFLOWER.

IVY-LEAVED BELL-FLOWER.—*Campanula hederacea.*

Class PENTANDRIA. *Order* MONOGYNIA. *Nat. Ord.* CAMPANULACEÆ. BELL-FLOWER TRIBE.

THIS remarkably elegant plant is not uncommon on wet heaths and by the side of streams in the south and west of England, and is very abundant in Cornwall. The small blossoms, on thread-like stems, are of pale, clear blue, and the leaves are shaped like those of the ivy, but of bright green. The plant is usually from four to six inches in height, but it sometimes climbs to twelve inches or more, when it can cling to the grass, amongst which it grows.

Of the eight wild species of Bell-flower, this is the smallest. Though the flowers of all are similar in form, and all of various shades of blue or purple, yet there is great variety in their size, from that of the little bell here represented, to the blossom of the Giant Bell-flower (*Campanula latifolia*), which is a frequent plant of Scotland, though not common in English woods. It is often reared, however, in our gardens, and, as well as some other species, is called Canterbury Bell. A learned writer on the usages of the Roman Catholic

Church supposes that this flower was so called
from its resemblance to the hand-bells which
were carried on a pole in the processions of
pilgrims to the shrine of Thomas à Becket in
those days when, as Chaucer writes—

"And specially from every shire's ende
    Of Englelond to Canterbury they wende,
    The holy blissful martyr for to seeke,
    That hem hath holpen whan that they were seke."

The Bell-flowers are still often used in
country places, as Clare says they were in his
boyhood's days :—

"When glowworm, found in lanes remote,
    Is murder'd for its shining coat,
    And put in flowers that Nature weaves
    With hollow shapes and silken leaves,
    Such as the Canterbury bell,
    Serving for lamp or lantern well."

But some insects are taught by their instinct
to turn these azure canopies to good account.
"The male of a little bee," say Kirby and
Spence, "a true Sybarite, dozes voluptuously
in the bells of different species of Campanula,
in which, indeed, we have often found other
insects asleep." Linnæus named a species
Florissimus because it loved to sleep in
flowers.

FIELD CHICKWEED.—*Cerastium arvense.*

Class DECANDRIA. *Order* PENTAGYNIA. *Nat. Ord.* CARYOPHYLLEÆ.
CHICKWEED TRIBE.

No less than eight species of these Mouse-ear Chickweeds grow wild in our native land, some of them being among the commonest plants, and flowering throughout the summer. The whole tribe are remarkably prolific, and supply in their seeds an abundant food for birds. They have all white blossoms, shaped much like those represented, but in most species they are much smaller.

The Field Chickweed thrives best on dry, sandy, and gravelly places, flowering in June. Its large blossoms of pure white are very pretty, and generally grow in the number of two or three at the top of the stalk. Its leaves are narrow and downy. The flower is described as very common in most counties of England, but the author does not find it so in Kent. It is rare in Scotland.

The Alpine Mouse-ear Chickweed (*Cerastium alpinum*), a low plant with white silky leaves and large white flowers, is very frequent in the Highlands of Scotland. Dr. Sutherland, in his recently published Journal of a Voyage to Baffin's Bay, which was made in

1851 and 1852, in search of Sir John Franklin,
describes this plant as growing on land little
removed from the sea and its immense icebergs.
" Button Point," he says, " looked as green as
any English meadow, and the grass upon it
was not one whit less luxuriant. The Foxtail
Grass and the Chickweed (*Cerastium alpinum*),
and hosts of other grasses and herbaceous
plants, grow among the bones of animals, and
are stimulated by the oil and animal matter
which they contain, and by the filth which is
inseparable from Esquimaux habitations, to a
degree of luxuriance which no one would be
willing to assign to the 73d deg. of N. lati-
tude." The Doctor adds, that the chubby
Esquimaux boy, filthy and greasy as he is,
takes his childish pastime, rolling on the
downy plots with which Nature provides him.

The name of this genus is from *ceras*, a horn,
from the shape of the seed-vessel in some of
the species ; and in several European countries
they are familiarly called by some word referring to this. Hence the Dutch term this
flower Hoornbloem, and the Germans, Das
Hornkraut. It is the Hornurt of the Danes,
and the Hornört of the Swedes.

FIELD GENTIAN.

FIELD GENTIAN.—*Gentiana campestris.*

Class PENTANDRIA. *Order* DIGYNIA. *Nat. Ord.* GENTIANEÆ.
GENTIAN TRIBE.

THIS blossom, of purplish lilac, is common on
dry hilly pastures from August till October.
The plant is about six or eight inches high, and
is very similar to another species called the
Autumnal Gentian (*Gentiana Amarella*), which
is to be found in similar places at the same
time. This latter kind may be known from
the species here figured by the circumstance
that its corolla is cut into five, instead of four,
segments. Both species have a delicate fringe
in the throat of the blossom, which is a most
beautiful object under a microscope.

Five species of Gentian belong to our wild
flowers. The Marsh Gentian (*Gentiana Pneu-
monanthe*), called Calathian Violet or Harvest
Bell, has deep blue blossoms marked with five
green stripes. It is a rare and a most
beautiful flower, growing in bogs. The Spring
Gentian too (*Gentiana verna*) is rare, but its
large single bell is of exquisite blue; and
hardly less beautiful or less infrequent is the
cluster of Small Alpine Gentian, (*Gentiana
alpina,*) which is found on the very summit
of the Highland mountains. The bitter principle contained in the Gentians has led to the

medicinal use of several species. The bitter Gentian of commerce is the *Gentiana lutea* of the Alps; but the root of our autumnal species is used in Russia; and the Purple Gentian (*Gentiana purpurea*) of Northern Europe is of as much power as the species usually employed. Sir J. E. Smith tells us that a ship from Norway once brought a quantity of this root to Edinburgh, where it was used with success, and found its way into the Pharmacopœia by the name of Cursuta,—a word which puzzled the etymologists, but which he believed to be a corruption of Skar-sote, "mountain soot," its Norway name.

Most of the Gentians are found where the country is hilly or mountainous. It was to some of this tribe that Coleridge alluded when he penned his magnificent poem to Chamouni :—

> " Who bade the sun
> Clothe you with rainbows? Who, with living flowers
> Of loveliest blue, spread garlands at your feet?
> God! let the torrents, like a shout of nations,
> Answer! and let the ice-plains echo, God!"

The beautiful large flower of *Gentiana major* grows in immense numbers within a few paces of the glaciers of Mont Blanc.

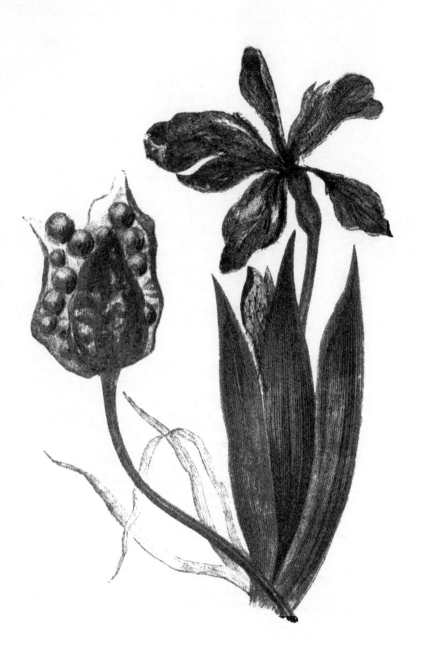

STINKING IRIS.

STINKING IRIS.—*Iris fœtidissima.*

Class TRIANDRIA. *Order* MONOGYNIA. *Nat. Ord.* IRIDACEÆ.
IRIS TRIBE.

THOUGH the familiar name of this flower is inelegant, yet, as might be inferred, it is expressive enough, for it has a most disagreeable odour, especially when bruised. The blossoms are of a dull leaden purple colour, and appear in June and August. The long, flat, sword-like leaves are so acrid, that they will, if bitten, produce a burning sensation in the tongue and lips, and are said even to loosen the teeth. This Iris is frequent in woods and hedges in the west and south of England, and is particularly abundant in Devonshire. The scarlet berries, when lying in their half-open capsule, are exceedingly beautiful; and they continue to ornament the plant through the dreariest season, or if gathered, form a handsome addition to the winter nosegay, as they will continue bright for some months.

This flower is sometimes called the Gladwyn Iris. In other places it is known as the Roast Beef Plant, on account of some supposed resemblance in its odour to that of this old

English dish. It shares with other kinds of Iris the name of *Flower de Luce*, but this should rather be applied to the lily. The introduction of this flower as a type in heraldry may be referred to the time of the Crusades, when Louis VII. of France chose it as his emblem. It appears, however, in the crown worn by Edward the Confessor, according t a coin engraved both in Speed and Camdem The *Fleur de Lis*, as its name imports, was evidently a lily; and the lily is represented in all ancient religious paintings of the Virgin, and in the old illuminated missals, either alone or in combination with the rose, as her invariable accompaniment. Our purple Iris is the only native species of this colour, though every variety of blue, as well as of yellow, is found in gardens. The only other wild Iris is the bright yellow flower of our streams, which has already been described. The name, which belonged to the heathen goddess of the rainbow, was given to the genus because of the various and beautiful hues of the blossoms.

WOOD SAGE.

WOOD SAGE.—*Teucrium Scorodonia.*

Class DIDYNAMIA.   *Order* GYMNOSPERMIA.   *Nat. Ord.* LABIATÆ.
LABIATE TRIBE.

THE wrinkled leaves of this plant are much like those of the true Sage (*Salvia*), but it has little other affinity with it. It is sometimes called Garlick Sage, because when bruised it has some slight odour of garlick, and in many places it is known as the Wood Germander. The spike of greenish yellow flowers, though not conspicuous, is pretty, and may be found from June to October, in woods, and on hedges, and other dry places.

No one would taste the bitter leaf of this plant without being at once reminded of the flavour of the hop. Every part of the Wood Sage has somewhat of this bitter principle; and in Sweden as well as in Jersey it is used instead of the hop in brewing. It is said that by its use the beer sooner becomes cleared than by the hop, but it darkens the colour of the liquor. In Jersey, when the usual beverage of cider has failed, the people make their home-brewed beer of the sage without any admixture of the hop, and they call the plant

by the name of *Ambrosie*. The synonym of
Ambrosia was one of its oldest names in our
own land, and apparently alludes to something
of a divine and undying nature which the plant
was thought to possess. Some English writers
on agriculture have recommended its cultivation, alleging that it is scarcely inferior to the
hop in its qualities, and that it would require
far less amount of expenditure than that plant.

Another native species is the Water Germander (*Teucrium Scordium*), a rare plant of
marshes, with purple flowers. It was, in early
times, regarded as a most valuable medicine
in the plague, and other pestilential disorders.
A third species, which is generally called the
Wall Germander (*Teucrium Chamœdrys*), is
rare also, and is indeed a doubtful native. It
has very small purple flowers.

The genus derives its name from Teucer,
who is said first to have used some of the
species medicinally. The Water Germander
is still employed as a tonic medicine in some
villages.

STRAWBERRY TREE.—*Arbutus Unedo.*

Class DECANDRIA. *Order* MONOGYNIA. *Nat. Ord.* ERICÆ.
HEATH TRIBE.

THIS beautiful tree, with its dark glossy evergreen leaves, is better known to most of us as a garden, than as a wild plant. Yet it is wild in the south of Ireland, and is abundant in woods at Mucross and Glengariff, near Bantry, as well as about the lakes of Killarney. Some writers think that it was introduced into this country from Spain or Italy, by the monks of Mucross Abbey, and doubtless we owe many plants to those who tended the gardens of the monasteries. No accurate account even of the trees or flowers then known in Britain has descended from those days, so that an uncertainty remains as to the origin of many now considered wild. The Arbutus is in blossom in September and October, its beautiful waxen bells, tinged with green, hanging beside the more brilliant ripening fruits, which are the produce of last year's flowers. The berry is scarlet, and much like a large strawberry, but it has not the rich flavour of that fruit; and many writers, as Sir J. E. Smith, describe it as

" ungrateful." Hence the name *Unedo*, "One I eat," is said to have originated from the circumstance that we should not, after taking the first, desire a second. When not too ripe, however, these fruits are agreeable to many palates, and boys gather them from the bushes about Killarney, and carry them in baskets for sale. Like the berries of the Mountain Ash, they have astringent properties, and a large number should not be eaten. In Spain both sugar and spirit are extracted from them.

The lakes of Killarney are rendered superior in picturesque beauty to the Scottish lakes by the great variety of tint in their foliage. Inglis says that few spots can show more loveliness in this respect, their hues varying from those of the brownish-green heather to the dark full green of the Arbutus.

The two species of Bear-berry belong to this genus, but they are not frequent plants. The Red Bear-berry (*Arbutus Uva-ursi*), is abundant on mountainous heaths in the North; and the Black Bear-berry (*Arbutus alpina*) is common on mountains in the north of Scotland.

COMMON HORNWORT.

COMMON HORNWORT.—*Ceratophyllum demersum.*

Class Monœcia. *Order* Polyandria. *Nat. Ord.* Ceratophylleæ. Hornwort Tribe.

Most lovers of the rural walk are lovers too of the streams, which, with " quiet tune," wander through the grass and sedges. Those who linger looking down into the waters, see plainly that they have a vegetation peculiar to themselves. Not to tell of showy flowers, as Water-lilies, and Flags, and Willow herbs, there are aquatic grasses, and sprays of small leaves, and masses of green threads which have no flowers, or flowers so small as not to be seen without close examination. The simplest form of the aquatic vegetation are those slimy masses which seem to have little right to one of their old names of Flowers of Heaven, and Falling Stars ; then come the Confervæ, many of them commonly known by the name of Crow-silk, and often forming tufts of brown or green silky jointed threads : while duckweeds, like little bright green leaves, and pond-weeds with foliage of dull olive, and star-worts, and feather-foils, and Hornwort, in their various forms, serve to render the pool as luxuriant as the meadow.

There are, in our streams, two species of Hornwort, very similar in their general appearance. The kind here represented is a very common plant in slow rivers and ditches. It is buoyed up entirely under water, and has long slender stems, around which the leaves grow in whorls, forming a more or less compact cone, the thread-like leaves being forked. The green flowers, which appear in July, are very small, and situated close to the stem. The fruit which succeeds them has two thorns near the base, and this is the chief distinction between the common Hornwort and the other species.

The Unarmed Hornwort (*Ceratophyllum submersum*) is less general in our fresh waters, but grows in ditches at the south and east of England. A species of Hornwort very similar to these is described as growing in the mineral thermal springs of Albano, where the temperature of the waters is 95° of Fahrenheit. The genus is named from two Greek words, signifying horn and leaf.

COMMON HOLLY.

COMMON HOLLY.—*Ilex Aquifolium.*

Class TETRANDRIA. *Order* TETRAGYNIA. *Nat. Ord.* ILICINEÆ.
HOLLY TRIBE.

THE Holly bough is too distinctly associated with some of the happiest days of childhood, to need description. Christmas-eve, with the Holly, and mistletoe, and ivy, brought in to deck the home—Christmas-day, with its red berries mingled with the ivy and yew to ornament the church, are seasons and circumstances not to be forgotten in life's later days; and the sight of these evergreens can recal to most some joyous remembrances, mingled, perhaps, with sad thoughts of some who hailed the festive seasons in other days, and whose voices shall gladden them no more. In many places our old Christmas usages are falling into disuse, and we can join in the lament of Clare :—

" Old customs, oh, I love the sound,
   However simple they may be?
Whate'er with time hath sanction found,
   Is welcome and is dear to me:

Pride grows above simplicity,
   And spurns them from her haughty mind;
And soon the poet's song will be
   The only refuge they can find."

Familiar as the red berries of the Holly are to us all, yet its flowers are less generally known. These grow closely round the stem in May and June, and are white, and thick, as if

cut out of wax. . The dark evergreen glossy
leaves are usually edged with sharp spines;
but the upper ones are often smooth, a
circumstance on which Southey founded his
beautiful little poem of the Holly-Tree. The
plant is very frequent in woods and hedges,
especially where the soil is of gravel; and
though it is of slow growth, yet it will, in the
course of time, form an excellent garden hedge.
All readers of John Evelyn's Diary will remember how much he lamented the destruction
of his beautiful Holly hedge, at Saye's Court,
by the Czar of Muscovy. This hedge was 400
feet in length, 9 feet high, and 5 feet thick; and
its owner might well mourn over its destruction.

In France, the young shoots of the Holly
are given as winter food to sheep and deer.
The wood is very white, and is used for painted
screens, and various articles wrought by the
turner. Birdlime is made from the bark. The
followers of Zoroaster believe that the sun never
shadows the Holly-tree, and the modern Parsees
are said to throw water, in which this plant is
infused, on the face of a newly-born infant.

PREMORAL SCABIOUS.

PREMORSE SCABIOUS.—*Scabiosa succisa.*

Class TETRANDRIA. *Order* MONOGYNIA. *Nat. Ord.* DIPSACEÆ. TEASEL TRIBE.

THIS pretty Scabious, with its purplish blue blossoms, forming nearly globose heads, is frequent on hills and commons, from August to October. As it grows so much on downs among the short grass, it is appropriately called by the Dutch, by a word signifying Turf-weed. Its old name of Devil's Bit Scabious, and its French synonym of Mors du Diable, were given on account of its root, which terminates abruptly. Our old herbalists knew it by this name. "Fabulous antiquity," says one of them, " (the monks and fryers, as I suppose, being the inventors of the fable,) said that the Deville, envying the good that this herb might do to mankind, bit away part of the root of it, and thereof came the name of Succisa and Deville's Bit."

We can remember having been sometimes disappointed on pulling up the root, at finding that it was not always thus premorse; but Dr. Drummond says that it is so only after the first year. He adds, that from this period the root becomes woody, dies, and, with the

exception of the upper part, decays, and that
this causes the eroded and bitten appearance;
while the new lateral branches, shooting out
from the part left, compensate for the want of
the main root. The author finds that in
specimens in which the root has the bitten
appearance, the fibres often end so abruptly,
that they too seem to have been bitten off.

We do not, in modern days, recognise any
great medicinal powers in this Scabious. The
root is astringent, and an infusion made with
it has a slightly bitter flavour. Linnæus tells
us that a yellow dye is obtained from the
leaves.

The flower, growing on a slender stem, a
foot, or sometimes two feet in height, is conspicuous on the open downs. The Rev. T. Bree
has found it with white blossoms, near Allesley,
in Warwickshire. The other species, the Small
Scabious (*Scabiosa columbaria*), blooms earlier,
and has a pale lilac flower, more resembling
that of the Knautia, already described. It
has also cut leaves, whereas those of the
premorse species are oblong and undivided.

COMMON NIPPLE-WORT.—*Lapsana communis.*

Class SYNGENESIA. *Order* ÆQUALIS.' *Nat. Ord.* COMPOSITÆ. COMPOUND FLOWERS.

THIS plant, which produces its numerous yellow starry flowers in July and August, may often, also, in a mild season, be found in bloom as late as December. It is not handsome, the blossoms being small and of a rather pale yellow, and the stalks long and straggling. The stem is about two or three feet high, and very much branched. The leaves are very thin, soft, and slightly hairy, and vary much in shape on the same specimen. Some are close to the stem, and heart-shaped at the base; others are long and slender; and those at the lower part of the plant are lyre-shaped.

This flower is very common in almost all hedges and waste places, but it is one which would oftener be called a weed than a flower. In some places it is known by the name of Succory Dock Cress, or Swine's Cress. The foliage is slightly pungent and bitter. It is said to be boiled for greens in some parts of England, but it is so inferior for this purpose to many other common wild plants, that it

is not likely to be generally used, the bitter
taste remaining even after boiling. It could
certainly only be eaten in spring, when the
shoots are young, and at this season the small
tops are flavoured something like the radish.
In Turkey they are sold in the markets for
salad, but climate probably modifies their
flavour.

The common species often grows as a weed
in gardens, but the only other British species
is more rare. This is the Dwarf Nipple-
Wort (*Lapsana pusilla*), which, though much
branched, is not above six or eight inches high.
It has small yellow flowers in June and July,
and is found in corn-fields or gravelly soils.
This kind is easily known from the other,
as well as from plants somewhat similar, by its
pipe-like stems, which are thickened and club-
shaped at the extremities. The larger species
is frequent throughout Europe. The Germans
call it Rainkohl; the Dutch, Lorchenboom;
and it is La Lampsane Commune of the French.
The scientific name is of Greek origin.

HEMP AGRIMONY.

HEMP AGRIMONY.—*Eupatorium Canna-binum.*

Class SYNGENESIA. *Order* ÆQUALIS. *Nat. Ord.* COMPOSITÆ. COMPOUND FLOWERS.

THOUGH the blossoms of the Hemp Agri-mony, forming clusters of purplish pink colour, are not showy, yet the plant is conspicuous on river sides by its large size, and by growing in great numbers together. It flowers in July and August, and is frequent in moist places. It has a slightly aromatic odour, and the branched reddish stems are three or four feet high, thickly studded with downy leaves. There is little variety in the tint of either leaf or flower, and neither has a bright or gay appearance.

This plant is bitter in flavour, and a decoc-tion of its root is an old medicine, and has some tonic virtues, but as an over-dose would be dangerous, it is to be regretted that it should be used as a village medicine. About the dykes of Holland it thrives very plentifully, and the Dutch, who call it Boelkenskruid, use it as a remedy for various disorders. It is likely that they act wisely in choosing it for this purpose, for as Dr. George Moore has re-marked, when speaking of tonics, " Bitters are more relished and more useful in marshy districts than in those more salubrious. The poor, on the coast of Sussex, use a strong infusion of that excellent bitter, the Lesser Centaury, with success in brow-ague, and the intermittent head-ache, so common among them." The author has also observed the good effects of this latter herb, when used medicinally, in some marshy districts of Kent.

Little importance can be attached to the opinion of the ancients respecting the remedial effects of plants, because they were influenced by many fanciful notions respecting them. The Hemp Agrimony, however, was prized in early times, and Pliny relates that it received its name Eupator, which was the surname of Mithridates, king of Pontus, because he first discovered its excellent qualities. It is said that if the plant be placed on shelves where bread is kept, it will preserve it from becoming mouldy.

The Common Hemp Agrimony is the only British plant of the genus. The Fever-wort of North America is the perfoliate species; and an American physician, some years since, published a dissertation on the medical virtues of this plant.

COMMON MARJORAM

COMMON MARJORAM.—*Origanum vulgare.*

Class DIDYNAMIA. *Order* GYMNOSPERMIA. *Nat. Ord.* LABIATÆ. LABIATE TRIBE.

THE pleasant odour of this plant, as well as the place of its growth, renders its scientific name very suitable. This is derived from the two Greek words, *Oros,* a mountain, and *ganos,* a joy; and a joy and delight it is to all who wander where it is plentiful. It is most abundant on hilly places and dry banks, especially such as have a chalk or limestone soil, where it grows to about the height of a foot or more. Its cluster of lilac, or purplish flowers, seem of a dark chocolate tint when only half expanded, in consequence of the numerous small leaves near the calyx, which the botanist terms bracts, and which are of this deep hue. It blooms in July and August, and it is often cultivated in kitchen gardens for domestic uses. An infusion of its leaves is in good repute as a tea in country places, and it is very aromatic and agreeable. In Sweden the Marjoram is used in beer to preserve it from becoming acid, and it is said to add to its intoxicating powers. The young tops dye woollen cloth of a purple colour.

The Marjoram was formerly much employed in medicine, and a strong essential oil exists in every part of the plant. The oil is very acrid and caustic, and a stimulating liniment is obtained from it, which is applied with good effect in cases of rheumatism. A small piece of wool dipped in the juice and put into the mouth, will often allay the toothache, and the whole plant is used in fomentations and medicated vapour baths.

The Common Marjoram is the only British species of the genus. It was formerly called Organy, and the long-disputed Oregon territory is said to have received its name from the prevalence of this plant there. It is common throughout Europe, and is very generally known by some name similar to our Marjoram. Forskahl says that its Arabic name is Maryamych, which the Arabs pronounce Marjamie.

WILD CARROT.—*Daucus Carota.*

Class PENTANDRIA. *Order* DIGYNIA. *Nat. Ord.* UMBELLIFERÆ.
UMBELLIFEROUS TRIBE.

THE Wild Carrot may be distinguished from other umbelliferous plants, by the purple or reddish-coloured flower which is found in the centre of the umbel, the whole of the central umbel being sometimes also of dark red or purple. The odour of the root too is exactly like that of the cultivated variety of our gardens, but it is either white or of a pale yellowish tinge. The leaves are handsome and graceful, and were formerly worn by ladies as feathers.

This plant is, in country places, commonly called Bird's Nest, because the stalks of the cluster, after flowering, stand upright and form a concave circle something like a nest. Though the bird makes no use of this, yet it seems that some of the insect race avail themselves of so good a place for their nightly repose, for Kirby and Spence mention their having found a curious and rare species of bee asleep in this sheltered dormitory.

The Wild Carrot is a very frequent plant on pastures and the borders of fields, and is the only British species, the kind called the Sea-side Carrot being most probably but a variety of it.   It is biennial, and is a tough-stemmed and somewhat bristly plant, usually rather more than a foot in height, and flowering in July and August, its white clusters having, in some specimens, a delicate tinge of pink. The seeds are bristly, and are strongly aromatic both in odour and flavour; and if placed in water they will impart to it a strong scent. They are considered carminative in their nature.   The Wild Carrot is generally found throughout Europe, and Sir T. L. Mitchell, when in Australia, saw a very similar species in abundance there, which he says he doubted not was a good vegetable; but his men, having suffered illness from eating a wild pea of that country, naturally enough declined trying the goodness of the carrot.

The Carrot derives its name, according to Theis, from *Car*, red, in Celtic; whence also come *Garance*, the French name for the red Madder roots, and our words carmine and carnation.

SPIKED PURPLE LOOSESTRIFE.
Lythrum Salicaria.

Class DODECANDRIA.   *Order* MONOGYNIA.   *Nat. Ord.* LYTHRARIÆ.
LOOSESTRIFE TRIBE.

THE Long Purples, as country people call the Loosestrife, are among the most conspicuous as well as the most richly-tinted of all the flowers which grow among the rushes and sedges of our streams.   Far away, and long ere any other blossom can be discerned, the tall rich spikes may be seen bowing slightly to the breeze which ripples the current, and colouring the landscape for miles along the margin of the waters.   The square stem is two, three, or sometimes even four feet high, and often more than a foot of its upper portion is during July and August crowded with the whorls of blossoms.

This flower is in some countries called Purple Grass, Willow-Strife, Purple Lythrum, or Willow Lythrum.   The French call it La Salicaire, and there is also a reference to the willow-shaped foliage, in the specific name of Salicaria, from Salix, a willow.   The long narrow leaves are either opposite on the stem, or form a whorl of three or more around it.

The colour of the flower is named from the Greek *lythron*, blood; and its hue is certainly much more like the ancient than the modern purple, having a rich crimson tint upon it.

This flower is common on moist lands, both in our own country and those of several other parts of Europe. It is frequent too in Australia, and Sir T. L. Mitchell says that it "raises its crimson spikes of flowers among the reeds of the Macquarie, as it does in England, on the banks of the Thames."

The only other species is a very different plant, and has less pretensions to beauty. It is the Hyssop-leaved Loosestrife (*Lythrum hyssopifolium*). It seldom exceeds three inches in height, has solitary purple blossoms and alternate leaves, and is rather a rare, or, at least, a local flower on lands occasionally inundated, or on which water has stood. It occurs chiefly in the eastern counties of England; its flowers are small and of dark purple, and it blooms in August.

COMMON GRASS OF PARNASSUS.
Parnassia palustris.

Class PENTANDRIA. *Order* TETRAGYNIA. *Nat. Ord.* HYPERICINEÆ.
ST. JOHN'S WORT TRIBE.

THOSE who are not great observers of plants, would, perhaps, smile at the predilection which the botanist feels for boggy land. But some of our most lovely plants grow there, and in such places, especially in the northern counties of England, and in Wales, we may find the Grass of Parnassus, with its delicate cream-coloured flowers, beautifully veined, and the leaves around the root on long stalks, while here and there a leaf clasps the flower-stem. The corolla has a singular appearance, from the fan-shaped scales, fringed with white hairs, which lie around the centre, and which are the nectaries, a little wax-like gland being at the tip of each hair.

Any one who looked at this plant would wonder why it should have received its English name of Grass, for it has no affinity whatever to the true grasses, with their hollow jointed stems, and green flowers. We can give no reason, unless it was meant to convey the idea that the plant was common as the very grass

itself on Mount Parnassus.   Gerarde is highly
indignant at its being so misnamed. "The
Grasse of Parnassus," he says, "hath heretofore
been described by blinde men; I do not meane
such as are blinde in their eyes, but in their
understandings, for if this plant be a kind of
grasse, then may the Butter-burre or Colte's-
foote be reckoned for grasses, as also all other
plants whatsoever." This flower was also in
former times called the Noble Liverwort, and
was, doubtless, like some others which bore
similar names, considered as a remedy in
disease. It grows on spongy bogs, in most
European countries, and its name generally has
some reference to the classic mount. Thus
the French term it Fleur de Parnassus; the
Dutch call it Parnaskruid; the Italians,
Spaniards, and Portuguese, all know it by the
name of Parnasia.

The stem of this flower is angular, and
usually about eight or ten inches high; but
Sir William Hooker found it growing with
perfect flowers, and the stem only about an
inch in height, at North Ronaldsha, in Orkney.
It is the only British species.

SD - #0037 - 030723 - C0 - 229/152/22 - PB - 9780282752934 - Gloss Lamination